Symbol	Meaning
Q	The set of rational numbers
$A \cup B$	The union of sets A and B
$A \cap B$	The intersection of sets A and B
(a, b)	The ordered pair a, b
$A \times B$	The Cartesian product of sets A and B
$a + b$	The sum of a and b
$a \cdot b, a \times b, (a)(b), ab$	The product of a and b
$(a, b, c, \ldots)$	The ordered set whose elements are $a, b, c, \ldots$
$<$	Is less than
$\not<$	Is not less than
$\leq$	Is less than or equal to
$>$	Is greater than
$\not>$	Is not greater than
$\geq$	Is greater than or equal to
$\sqrt{x}$	The positive square root of x
$\equiv$	Is congruent to
$[a]$	The equivalence class of a
$P(n, r)$	The number of permutations of elements taken r at a time
$C(n, r)$	The number of combinations of elements taken r at a time
$P(A)$	The probability of A
$P(\overline{A})$	The probability of not-A

Mathematics

A Liberal Arts Approach

MATHEMATICS
A LIBERAL ARTS APPROACH

MALCOLM GRAHAM
University of Nevada, Las Vegas

HARCOURT BRACE JOVANOVICH, INC.

New York　　*Chicago*　　*San Francisco*　　*Atlanta*

ISBN: 0-15-555235-X

Library of Congress Catalog Card Number: 73-918

Printed in the United States of America

COVER   Courtesy, Dover Pictorial Archives Series.

PHOTO CREDITS   Photos on pages 240, 241, 247, and 249 courtesy of International Business Machines Corporation. Photos on pages 245 and 246 courtesy of Burroughs Corporation.

PREFACE

This book was written primarily for college students who are not majoring in mathematics. Prospective teachers of elementary and secondary school mathematics will also find the material useful. The principal objectives of the book are to develop an appreciation for mathematics as a creative art and science, to provide an insight into the methods of reasoning used in mathematics, and to show the role that mathematics has played in the history of mankind.

Various topics and exercises are presented to exemplify the nature and spirit of mathematics. However, the reader is not overburdened with problems that involve lengthy manipulative procedures. Emphasis is placed on mathematical concepts, general principles, and strategies, rather than on manipulation. The book may be studied with only a knowledge of arithmetic; occasionally some background in algebra and geometry is helpful, but certainly not essential. Sufficient material is provided for a three or four semester-hour course, depending on the background of the student and the depth of coverage desired by the instructor.

Although the chapters are largely independent of each other, there is a pattern in their arrangement. Logic and the nature of proof are considered first to develop a feeling for the methods of thinking involved in the development of mathematical concepts. The concept of sets and operations on sets leads naturally into the development of numbers and number systems. If the students have been exposed to sets and number systems, Chapters 2, 3, 4, and 5 may be covered briefly. Many of the topics in Chapters 6 through 10 have been developed in recent times, and in some cases their study brings the student to the frontier of mathematical knowledge without requiring a great deal of background.

The instructor who wishes to cover the first part of the text in depth should be aware that the last four chapters (7, 8, 9, and 10) are independent of each other and may be studied in any order or omitted, as necessary.

There are 53 sets of exercises in the text as well as 9 review sets at the ends of the chapters. (Chapter 10 has no review exercises.) Answers are supplied for all exercises except the review exercises and those problems with similar parts, in which case only the odd parts are answered. Answers to exercises not given in the text are available to the instructor.

In the writing of this text many helpful suggestions were offered by various mathematicians and mathematics educators. In particular, I am grateful for the constructive criticisms of Professor Leon Faure, College of San Mateo; Professor Charles L. Murray, Chabot College; and my wife Carolyn L. Graham, Clark County School District. Special thanks go to several students at the University of Nevada, Las Vegas, who assisted in working out solutions to the exercises, and to Diana Stanga for her excellent typing of the manuscript.

Malcolm Graham

CONTENTS

3 Operations on Sets, 53

4 Operations on Numbers and the Use of Number Bases, 73

5 The Evolution of Numbers and Number Systems, 101

The Theory of Numbers, 141

Finite Mathematical Systems and the Congruence Relation, 159

Probability, 177

Mathematics

A Liberal Arts Approach

Auguste Herbin, *Composition on the Name "Rose"*
1947. Oil on canvas. 32 x 25⅝".
The Solomon R. Guggenheim Museum, New York

1 LOGIC AND THE NATURE OF PROOF

1.1 Introduction

Mathematics is an area of study in which a high degree of creativity may be expressed. Mathematics is created by man, and the rate at which it is being created has been steadily accelerating in recent times. Just as in many other fields, however, the beginner must first master the simpler elements before he attempts the harder ones; the student of mathematics, therefore, must learn certain basic skills and concepts before he can add new knowledge to what has already been developed. How, then, can mathematics be a creative subject for the undergraduate college student? He may solve problems independently that have already been solved by someone else, pose new problems, and perhaps attempt to solve some of the classical problems. The satisfaction of climbing a mountain that you have never climbed is only slightly diminished by the fact that others have climbed it. Many people enjoy working a crossword puzzle knowing full well that the answer is provided on the next page. Solving a problem in mathematics without assistance can be equally rewarding.

A mathematician attempts to prove certain statements concerning numbers, points, lines, planes, and so forth. Frequently we are interested in how these results might apply to situations in other areas of study, such

3

as physics, chemistry, biology, engineering, architecture, art, the social sciences, business, and economics. Even if the reader's primary interest is not in mathematics per se, logic and the nature of mathematical proof can be important to him. In this chapter we will discuss several types of arguments commonly used in mathematical proofs.

The nature of a valid argument can be better understood if we first develop some preliminary concepts of logic. If the study of this chapter seems a bit more difficult than some of the following chapters, you might find it reassuring to know that symbolic (or mathematical) logic is not considered to be of a trivial nature. Although even in the time of Aristotle (384–322 B.C.) men were interested in the nature of argument, the first significant work in symbolic logic was relatively recent. The British mathematician George Boole (1815–1864) laid the foundation for contemporary studies in logic. He wrote a pamphlet called *The Mathematical Analysis of Logic* (1847) and later had a more mature statement of his views published under the title *An Investigation of Thought, on which are founded the Mathematical Theories of Logic and Probability* (1854). Boole believed that the theories of logic could be applied to an almost unlimited variety of statements even though he expressed these theories in mathematical terms. The branch of mathematics known as Boolean algebra *set theory* is a memorial to his contributions, and the many applications of Boolean algebra to such things as switching gear for telephones and the design of electronic computers testify to the value of his theories. Bertrand Russell said, "Pure mathematics was discovered by Boole in a work he called *The Laws of Thought*."

Today the study of symbolic logic is considered indispensable for understanding the foundations of mathematics and the nature of proof. Therefore, let us examine a few introductory concepts of logic.

1.2   Statements

We begin our study of logic by defining a few terms. As you know, a simple sentence usually contains a subject and a predicate. We will assume the usual definition of a sentence and define the word *statement* as follows.

DEFINITION 1.2.1. A statement is a sentence that is true or false, but not both.

Examples of statements are

(a) A triangle has three sides.
(b) Harrisburg is the capital of Pennsylvania.
(c) $2 + 3 = 9$.

(d) Wednesday is the first day of the week.

(e) Carolyn went to the movies.

Whether a statement is true or false is referred to as its **truth value.**
Statements (a) and (b) above are true, while statements (c) and (d) are false.
Although we may not know whether a statement such as (e) is true or false,
the important thing is that it must be either true or false, but not both.
In some particular instance, of course, we may wish to know which is the
case. In fact, much of our discussion will be concerned with the conditions
under which certain kinds of statements are either true or false.

Expressions such as "open the window" or "row your boat" are *not*
statements since they are neither true nor false. Hence they will not be
considered in our discussion.

1.3 Conjunction, Disjunction, and Negation

The examples of statements that have been given so far are *simple* statements, each of which has a single subject and predicate. A *compound*
statement may be formed by joining two simple statements with a *connective.* (Later, we will extend this concept by joining more than two simple
statements with several connectives.) Some connectives that are commonly
used in logic and mathematics are listed below. We will discuss the first
three in this section of the text and the last two in the following section.
With the exception of the connective *not*, each of the connectives listed
below is used to form a compound statement. Note that P and Q each
represent simple statements.

Connective	Symbol for Connective	Statement	Symbolized Statement	Name of Statement
And	$\wedge$	P and Q	$P \wedge Q$	Conjunction
Or	$\vee$	P or Q	$P \vee Q$	Disjunction
Not	$\sim$	not P	$\sim P$	Negation
If–then	$\rightarrow$	If P, then Q	$P \rightarrow Q$	Implication
If and only if	$\leftrightarrow$	P if and only if Q	$P \leftrightarrow Q$	Equivalence

Whether a compound statement is true or false depends upon the
truth value of each of its component simple statements and the connective
used. If the connective *and* is used, for example, the compound statement
$P \wedge Q$ is defined to be true only if *both P and Q* are true. The statement
$P \wedge Q$ is called a **conjunction.** The four possible types of conjunctions are

determined by considering all combinations of truth values for P and Q as summarized in the **truth table** of Figure 1.3.1.

FIGURE 1.3.1

P	Q	$P \wedge Q$
T	T	T
T	F	F
F	T	F
F	F	F

The four possible types of conjunctions are illustrated below.

	$P \wedge Q$			Truth Value of $P \wedge Q$
Joe went skiing (T)	*and*	Paul went sailing. (T)		T
Joe went skiing (T)	*and*	Paul went sailing. (F)		F
Joe went skiing (F)	*and*	Paul went sailing. (T)		F
Joe went skiing (F)	*and*	Paul went sailing. (F)		F

Next we will consider the connective *or*. In everyday usage, the word *or* has two possible meanings. In one situation, P or Q might mean either P or Q but not both (the *exclusive* usage of *or*); in another situation it might mean either P or Q or both (the *inclusive* usage of *or*). For example, "I will go to the movies or stay home" would mean that I will either go to the movies or stay home but not do both, whereas "Do you like golf or tennis?" is probably asking whether you like either golf or tennis or both. To avoid the possibility of misinterpretation, we will define P or Q (written $P \vee Q$) to mean *either P or Q or both*. Note that this is the inclusive usage of or. Thus, the only way $P \vee Q$ can be false is if both P and Q are false. The statement $P \vee Q$ is called a **disjunction**. The four possible types of disjunctions are shown in the truth table of Figure 1.3.2.

FIGURE 1.3.2

P	Q	$P \vee Q$
T	T	T
T	F	T
F	T	T
F	F	F

The following example illustrates the four possible types of disjunctions.

		$P \vee Q$			Truth Value of $P \vee Q$
Mary likes golf (T)	or	Mary likes tennis. (T)			T
Mary likes golf (T)	or	Mary likes tennis. (F)			T
Mary likes golf (F)	or	Mary likes tennis. (T)			T
Mary likes golf (F)	or	Mary likes tennis. (F)			F

It is very simple to analyze the connective *not* (symbolized $\sim$). If statement P is true, then $\sim P$ is false; if P is false, then $\sim P$ is true. *Not* P ($\sim P$) is called the **negation** of P. See Figure 1.3.3 for the truth table of the connective *not*.

FIGURE 1.3.3

P	$\sim P$
T	F
F	T

Recall that each of the connectives *and* and *or* was used to connect two simple statements to produce a compound statement. However, the connective *not* is connected to only one statement and hence does not produce a compound statement. For this reason, *not* is sometimes called a **unary connective** while *and* and *or* are called **binary connectives**.

Example 1. Negate the statement, "Fish swim." (A true statement.)

Solution: Fish do not swim. (A false statement.)

Example 2. Negate the statement, "Bats are not mammals." (False.)

Solution: Bats are mammals. (True.)

Example 3. Negate the statement, "No mammals live in the water." (False.)

Solution: Some mammals live in the water. (True.)

One must be extremely careful when negating statements involving expressions such as *some, all, many, every,* and *none;* such expressions are called **quantifiers**. Remember that if a statement is true, its negation must be false; and if a statement is false, its negation must be true. Consider the true statement: *Some mammals lay eggs.* (The duckbilled platypus lays eggs.) *Some mammals lay eggs* means that *at least one* mammal lays eggs, but it does not preclude the possibility that all mammals lay eggs or that some do not lay eggs. Hence the statements *some mammals do not lay eggs* and *all mammals lay eggs* are **not** negations of the original statement. A suitable negation would be: *No mammals lay eggs* or *It is not the case that some mammals lay eggs.*

More complex statements can be made by using various combinations of simple statements and connectives. Consider the conjunction $P \wedge \sim Q$ for example. The truth value of this compound statement will depend, of course, on the truth values of P and of $\sim Q$. The truth value of $\sim Q$ will in turn depend on the truth value of Q. One of the easiest ways to make a truth table for $P \wedge \sim Q$ is to first have columns for P and Q as shown in Figure 1.3.4. It is easy to construct a column for $\sim Q$ from the Q column since if Q is true, $\sim Q$ is false and if Q is false, $\sim Q$ is true. Then, using the P and $\sim Q$ columns, the truth values of $P \wedge \sim Q$ are readily determined since $P \wedge \sim Q$ is true only when both P and $\sim Q$ are true.

FIGURE 1.3.4

P	Q	$\sim Q$	$P \wedge \sim Q$
T	T	F	F
T	F	T	T
F	T	F	F
F	F	T	F

Example 4.   Prepare a truth table for $\sim(\sim P \vee \sim Q)$. (Remember a disjunction is false only if both of its simple statements are false.)

Solution: See Figure 1.3.5.

FIGURE 1.3.5

P	Q	$\sim P$	$\sim Q$	$\sim P \vee \sim Q$	$\sim(\sim P \vee \sim Q)$
T	T	F	F	F	T
T	F	F	T	T	F
F	T	T	F	T	F
F	F	T	T	T	F

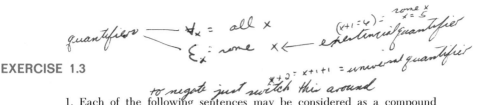

quantifiers — $\forall_x$ = all x

$\exists_x$: some x ← existential quantifier

$(x+1=6)$: some x, $x = 5$

$(x+3) = x+1+1 =$ universal quantifier

EXERCISE 1.3

to negate just switch this around $x+3 = x+1+1$

1. Each of the following sentences may be considered as a compound statement. Write the simple components of each.
   (a) Today is Monday or today is Wednesday.
   -(b) Joyce and Bill went to the movies.
   (c) Either you or I will pay the rent.
   -(d) It is both free and enjoyable.

2. Construct truth tables for $P \wedge Q$ and $P \vee Q$. Use these truth tables to label each of the following statements either true or false.
   (a) Birds fly and elephants fly.
   ────(b) $(4 + 13 = 17) \vee (2 + 4 = 23)$.
   -(c) $(6 + 12 = 18) \wedge (8 - 3 = 5)$.
   .(d) $(2 + 3 = 8) \vee (9 - 3 = 8)$.

3. Explain how the connective *not* differs from *and* and *or*. when we have not, it is a simple statement

- 4. (a) Make a truth table having columns for P, $\sim P$, and $\sim\sim P$.
   (b) How do the truth values of P and $\sim\sim P$ compare?
   (c) Predict the truth values for $\sim\sim\sim P$ as compared with those which have already been determined. Verify your hypothesis by adding another column to the truth table of part (a).

All (5.) Write the negation of each of the following:
   (a) Joyce is not tall.
   (b) John is always cheerful.
   (c) Some mammals can swim.
   (d) It is not the case that fish do not swim.

6. Letting H stand for "Henry is tall" and D stand for "Diana is tall," write each of the following in symbolic form.
   -(a) Henry is tall and Diana is short.
   (b) Diana is short and Henry is not tall.
   ~( ) →(c) It is not the case that both Henry and Diana are short.
   (d) Henry and Diana are both tall or Diana is short.

7. Make a truth table for each of the following:
   (a) $\sim P \vee Q$ (b) $\sim P \wedge \sim Q$
   (c) $\sim(P \vee Q)$ (d) $\sim(P \wedge Q)$
   -(e) $(P \wedge Q) \vee P$ (f) $\sim P \wedge (P \vee Q)$

8. Let G stand for "George likes poetry" and R stand for "Robert likes poetry."
   (a) In symbolic form write the statement "It is not the case that both George and Robert dislike poetry."
   -(b) Construct a truth table for this statement.
   (c) State in words the conditions under which the statement is true.

1.4 Implication and Equivalence

Many compound statements are of the form: *If P, then Q*. For example, we might say, "If rain falls, then the grass will grow." Such a statement may be referred to as an *if-then* statement and is called an implication or a conditional. We abbreviate an implication by writing $P \rightarrow Q$, which is read, "If P, then Q" or "P implies Q."

The same implication may be stated in many different ways. For example, the following are equivalent to each other:

(a) If rain falls, then the grass will grow.
(b) If rain falls, the grass will grow.
(c) The grass will grow if rain falls.
(d) Rain falls implies the grass will grow.
(e) (Rain falls.) $\rightarrow$ (The grass will grow.)

Determining whether an implication is true or false is somewhat more involved than determining the truth value of a conjunction or a disjunction. For this reason, we will first consider an example.

Suppose a man tells his wife, "If I get a salary increase, then I will buy you a new car." Under what conditions will his wife consider that he has made a true statement (kept his promise)? There are four possible situations:

(a) He gets a salary increase, and he buys his wife a new car. (Promise is kept. Original statement is *true*.)
(b) He gets a salary increase, and he does not buy his wife a new car. (Promise is broken. Original statement is *false*.)
(c) He does not get a salary increase, and he buys his wife a new car. (Promise is kept. Original statement is *true*.)
(d) He does not get a salary increase, and he does not buy his wife a new car. (Promise is kept. Original statement is *true*.)

The above statements are generalized in the truth table of Figure 1.4.1. Note that in this case P represents the statement "I get a salary increase," Q is the statement "I will buy you a new car," and $P \rightarrow Q$ represents the entire compound statement "If I get a salary increase, then I will buy you a new car." The most interesting situation is that when P is false, $P \rightarrow Q$ is true whether or not Q is true. It is sometimes difficult to accept the logical truth of statements such as "If $2 + 4 = 100$, then fish can swim," and "If $2 + 4 = 100$, then fish cannot swim."

FIGURE 1.4.1

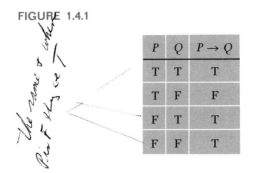

P	Q	P → Q
T	T	T
T	F	F
F	T	T
F	F	T

We can see from Figure 1.4.1 that $P \to Q$ is false only when P is true and Q is false. Similarly, $Q \to P$ is false only when Q is true and P is false. Hence we can easily expand the truth table to include $Q \to P$ as shown in Figure 1.4.2.

FIGURE 1.4.2

P	Q	P → Q	Q → P	P ↔ Q
T	T	T	T	T
T	F	F	T	F
F	T	T	F	F
F	F	T	T	T

In studying Figure 1.4.2, the reader will note that when $P \to Q$ is true, $Q \to P$ may or may not be true. This means that a statement such as "If it is 110° in the shade, then it is hot" may be true, while the statement "If it is hot, then it is 110° in the shade" may or may not be true.

Figure 1.4.2 also shows a column for $P \leftrightarrow Q$, which means the same as $(P \to Q) \wedge (Q \to P)$. This statement, of course, is a conjunction and is true only if both $P \to Q$ and $Q \to P$ are true. Using this information, we can see that the last column of Figure 1.4.2 is correct. Rather than reading $P \leftrightarrow Q$ as "If P then Q and if Q then P," we shorten it to read "P *if and only if* Q," or "P *is equivalent to* Q." The statement "P if and only if Q" is called an **equivalence** (or **biconditional**). Note, however, that the equivalence $P \leftrightarrow Q$ is true only when P and Q have the same truth values; that is, when P and Q are both true or both false.

Example 1. A statement such as "I will buy my wife a new car *if and only if* I get a salary increase" is true in only two cases. This can be verified from the truth table of Figure 1.4.2 which shows that $P \leftrightarrow Q$ is true only when P and Q are both true or when P and Q are both false. Hence, if I get a salary increase and I buy my wife a new car, I will have made a true statement; or, if I do not get a salary increase and I do not buy my wife a new car, I will have once again made a true statement. (If I do not get a salary increase and buy my wife a new car, I will have made a false statement—but my wife will forgive me! Will my wife forgive me if I get a salary increase and do not buy her a new car?)

Example 2.   Construct a truth table for $\sim P \to Q$.

Solution: See Figure 1.4.3.

FIGURE 1.4.3

P	$\sim P$	Q	$\sim P \to Q$
T	F	T	T
T	F	F	T
F	T	T	T
F	T	F	F

Example 3.   Construct a truth table for $(P \wedge Q) \to \sim P$.

Solution: See Figure 1.4.4.

FIGURE 1.4.4

P	Q	$P \wedge Q$	$\sim P$	$(P \wedge Q) \to \sim P$
T	T	T	F	F
T	F	F	F	T
F	T	F	T	T
F	F	F	T	T

It is easy to see that infinitely many statements can be made with varying degrees of complexity according to the number of simple statements and connectives used. Whether a statement such as $(\sim P \wedge Q) \vee (P \leftrightarrow Q)$ is true or false depends only on the truth values of P and Q. Since the truth values of such a complex statement are usually not immediately apparent, a truth table may be used to clarify the situation. Once the technique of making truth tables is understood, truth tables for complicated expressions are not particularly difficult to prepare—they merely take longer and the possibility of errors increases. From Figure 1.4.5 we can see immediately that $(\sim P \wedge Q) \vee (P \leftrightarrow Q)$ is false only when P is true and Q is false.

FIGURE 1.4.5

P	Q	$\sim P$	$\sim P \wedge Q$	$P \leftrightarrow Q$	$(\sim P \wedge Q) \vee (P \leftrightarrow Q)$
T	T	F	F	T	T
T	F	F	F	F	F
F	T	T	T	F	T
F	F	T	F	T	T

EXERCISE 1.4

1. Prepare a truth table with the following column headings: P, Q, $P \rightarrow Q$, $Q \rightarrow P$, and $P \leftrightarrow Q$.

2. Let M represent the statement "Mary is young" and J represent the statement "John is young," and write each of the following statements in symbolic form.
   (a) Mary is young and John is young.
   (b) John is not young and Mary is young.
   (c) Mary is young or John is not young.
   (d) If Mary is young, then John is young.
   (e) Mary is not young if John is not young.
   (f) Mary is young if and only if John is young.
   (g) John is young if and only if Mary is not young.
   (h) John is young implies Mary is young.

3. Make truth tables for:
   (a) $P \rightarrow \sim Q$ (b) $\sim P \rightarrow Q$
   (c) $(P \vee Q) \rightarrow Q$ (d) $Q \leftrightarrow (P \vee Q)$
   (e) $(P \wedge Q) \rightarrow (P \vee Q)$ (f) $(P \vee Q) \leftrightarrow (P \wedge Q)$

4. (a) Using M for "I will see a movie" and R for "It rains," make a truth table for the statement "I will see a movie if and only if it does *not* rain."
   (b) From the truth table of part (a) determine the conditions under which the given compound statement is true.

5. (a) Under what conditions do you believe $(P \vee \sim P) \leftrightarrow P$ will be true? (Do not use a truth table.)

   (b) Recalling that P and $\sim P$ *cannot* have the same truth values, make a truth table for $(P \vee \sim P) \leftrightarrow P$ and verify or refute your hypothesis for part (a).

6. (a) Make a truth table for $P \rightarrow \sim P$.
   (b) Using the truth table of part (a), what can be said about a person's statements if every statement he makes implies its own negation?

7. By constructing an appropriate truth table, prove that a statement is never equivalent to its own negation.

1.5 The Converse, Inverse, and Contrapositive of an Implication

In an implication "If P, then Q," symbolized by $P \rightarrow Q$, the statement P is called the **premise** or **hypothesis** and Q is called the **conclusion**. If the premise and conclusion of a statement of implication are interchanged, the new implication is called the **converse** of the first. The converse of $P \rightarrow Q$ is $Q \rightarrow P$. Hence, in considering two implications such as $P \rightarrow Q$ and $Q \rightarrow P$, we see that each is the converse of the other.

Other statements related to a given implication $P \rightarrow Q$ may involve the negations of P and Q. The **inverse** and **contrapositive** are two such statements. A summary of these four related statements is shown below.

Statement	$P \rightarrow Q$	If P, then Q
Converse	$Q \rightarrow P$	If Q, then P
Inverse	$\sim P \rightarrow \sim Q$	If *not* P, then *not* Q
Contrapositive	$\sim Q \rightarrow \sim P$	If *not* Q, then *not* P

Example 1. Write the converse, inverse, and contrapositive of the statement, "If you are a resident of New Jersey, then you are a resident of the U.S.A."

Solution:

   Converse: If you are a resident of the U.S.A., then you are a resident of New Jersey.

   Inverse: If you are not a resident of New Jersey, then you are not a resident of the U.S.A.

   Contrapositive: If you are not a resident of the U.S.A., then you are not a resident of New Jersey.

In studying Example 1, we see that the converse and the inverse do not in every case have the same truth values as the original statement. (See

Figure 1.5.1.) The contrapositive, however, does have exactly the same truth values as the original statement, and we say that the two statements are *logically equivalent*. Any two statements are said to be **logically equivalent** if they have the same truth values. Furthermore, if any two statements P and Q have the same truth values, then $P \leftrightarrow Q$ will always be true.

Referring again to Figure 1.5.1, note that although the truth values of the converse and the inverse are different from the truth values of the

FIGURE 1.5.1

same

P	Q	$\sim P$	$\sim Q$	Statement: $P \rightarrow Q$	Converse: $Q \rightarrow P$	Inverse: $\sim P \rightarrow \sim Q$	Contrapositive: $\sim Q \rightarrow \sim P$
T	T	F	F	T	T	T	T
T	F	F	T	F	T	T	F
F	T	T	F	T	F	F	T
F	F	T	T	T	T	T	T

same

original statement, they are identical to each other, and hence the converse and inverse are logically equivalent to each other. (Upon close examination, we see that this is to be expected since the inverse is simply the contrapositive of the converse!)

Many people are inclined to believe that if a statement is true, its converse and inverse will also be true. The truth table of Figure 1.5.1, however, shows us that if a statement is true, its converse and inverse may or may not be true.

Example 2. Consider the true statement, "If you are a woman, then you are a human being." The converse, "If you are a human being, then you are a woman," may or may not be true. (You might be a man.) Similarly, the inverse, "If you are not a woman, then you are not a human being," may or may not be true. (Suppose you are a man?) The contrapositive, however, is equivalent to the original statement and is true or false accordingly. The original statement was true, and we see it is also true that "If you are not a human being, then you are not a woman."

EXERCISE 1.5

1. Make a truth table for a statement of implication, its converse, its inverse, and its contrapositive. Compare your table with Figure 1.5.1. (You may wish to refer to this table in solving some of the following problems.)

2. State the converse, inverse, and contrapositive of each of the following statements of implication.
   (a) If you have long legs, then you can run fast.
   (b) If the car is green, then the car is not mine.

3. For which of the following statements is the converse necessarily true?
   (a) If Albany is the capital of New York, then Houston is the capital of California.
   (b) If you are John's sister, then John is your brother.
   (c) If the big Indian is the little Indian's father, then the little Indian is the big Indian's son.
   (d) If the animal is a mammal, then the animal is a horse.

4. Can $P \rightarrow Q$ and $Q \rightarrow P$ both be false? Explain.

5. (a) Write a true statement of implication that has a true converse.
   (b) Write a true statement of implication that has a false converse.
   (c) Write a false statement of implication that has a true converse.
   (d) Write a false statement of implication that has a false converse.

6. (a) Suppose Alice says, "If I do my homework, I will watch television." May we logically conclude that if Alice does not do her homework, she will not watch television? Explain.
   (b) Suppose Edward says, "If my name is in that book, then it belongs to me." May we conclude that if the book does not belong to Edward, then his name is not in it? Explain.

7. (a) What do we mean when we say that two statements are logically equivalent?
   (b) Using E for "I will eat" and S for "I will starve," construct a truth table to show that "I will not eat or I will not starve" is equivalent to "If I will eat, then I will not starve."

8. (a) Using P for "You will pass" and S for "You study," make a truth table for "You will pass if and only if you study."
   (b) Make a truth table for "You will not pass if and only if you do not study."
   (c) How do the truth tables of parts (a) and (b) compare?
   (d) Are the statements of parts (a) and (b) logically equivalent? Explain.

9. Augustus De Morgan (1806–1871) was a British mathematician. The following equivalences are known as De Morgan's Laws. Use truth tables to show that De Morgan's Laws are always true.
   (a) $\sim(P \wedge Q) \leftrightarrow (\sim P \vee \sim Q)$
   (b) $\sim(P \vee Q) \leftrightarrow (\sim P \wedge \sim Q)$

1.6   Types of Valid Arguments

A tautology is a statement that is always true regardless of the truth values of its component parts. "He is guilty (G) or not guilty $(\sim G)$" is a simple example. A truth table for $G \vee \sim G$ will immediately reveal that such a statement is always true and is, therefore, a tautology. However, since we will be discussing types of *valid arguments*, our interest will be almost exclusively in statements of implication—particularly those statements of implication that are always true (tautological implications).

The purpose of mathematical proofs and other types of valid arguments usually is to show that certain statements, called the **premises**, imply another statement, called the **conclusion**. In symbolic form,

$$(P_1 \wedge P_2 \wedge P_3 \wedge \ldots \wedge P_n) \rightarrow Q.$$

If the implication

$$(P_1 \wedge P_2 \wedge P_3 \wedge \ldots \wedge P_n) \rightarrow Q$$

is true regardless of the truth values of its component statements, it is a tautology and may be used in a valid argument. We are especially interested, however, in the fact that if the premises $P_1 \wedge P_2 \wedge P_3 \wedge \ldots \wedge P_n$ of a tautological implication are true, then Q *must* be true since a true implication $P \rightarrow Q$ cannot have a true premise P and a false conclusion Q.

For example, consider the following argument:

Premises: (1) If you water the lawn,
                       then the grass will grow.
                   (2) You water the lawn.

Conclusion: The grass will grow.

Observe that the above argument appears to be valid whether it is true or false that you water the lawn or that the grass will grow. But, *if* the premises are true, *then* the conclusion is true. The above argument has the general form:

Premises: (1) $P \rightarrow Q$
                   (2) P

Conclusion: $\therefore Q$ (The symbol $\therefore$ means *therefore*.)

This type of argument is called **modus ponens** or the **law of detachment**. If $P \rightarrow Q$ is true and P is true, we can "detach" Q and state that it is true. The justification of modus ponens, of course, depends upon whether or not $[(P \rightarrow Q) \wedge P] \rightarrow Q$ is a tautology (always true). Since this implication is always true, as shown in the truth table of Figure 1.6.1, modus ponens constitutes a *valid* argument.

FIGURE 1.6.1

P	Q	$P \rightarrow Q$	$(P \rightarrow Q) \wedge P$	$[(P \rightarrow Q) \wedge P] \rightarrow Q$
T	T	T	T	T
T	F	F	F	T
F	T	T	F	T
F	F	T	F	T

Truth Table for Modus Ponens
(Law of Detachment)

It should be emphasized that a valid argument does not depend on whether its premises are true or false, but rather on its structure. Hence Figure 1.6.1 shows that modus ponens can be used in a valid argument since $[(P \rightarrow Q) \wedge P] \rightarrow Q$ is true regardless of the truth values of its component statements. The following is a valid argument using modus ponens, even though some of the statements may be false.

If you study this book, you will be 20 feet tall.
You study this book.

Therefore, you will be 20 feet tall.

If the premises in the above argument were true, then would the conclusion be true? The answer, of course, is "yes" since it is a valid argument. However, since at least one of the premises may be false, we have no reason to believe that the conclusion is true even though the argument is valid.

Example 1. Given the statements: (P) You live in Los Angeles, and (Q) You live in California, consider the following arguments:
(a) Since $[(P \rightarrow Q) \wedge P] \rightarrow Q$ is a tautology, a *valid* argument using *modus ponens* is:

Premises:	(1) If you live in Los Angeles, then you live in California.	(1) $P \rightarrow Q$
	(2) You live in Los Angeles.	(2) P
Conclusion:	You live in California.	$\therefore Q$

(b) Since a truth table will show that $[(P \rightarrow Q) \wedge Q] \rightarrow P$ is *not* a tautology, an *invalid* argument is

Premises:	(1) If you live in Los Angeles, then you live in California.	(1) $P \rightarrow Q$
	(2) You live in California.	(2) Q
Conclusion:	You live in Los Angeles.	$\therefore P$

(c) A truth table will show that $[(P \rightarrow Q) \wedge (\sim P)] \rightarrow \sim Q$ is *not* a tautology. Hence, another *invalid* argument is

Premises:	(1) If you live in Los Angeles then you live in California.	(1) $P \rightarrow Q$
	(2) You do not live in Los Angeles.	(2) $\sim P$
Conclusion:	You do not live in California	$\therefore \sim Q$

Another *valid* type of argument is modus tollens. Referring to Example 1, consider the following argument:

Premises:	(1) If you live in Los Angeles, then you live in California.	(1) $P \rightarrow Q$
	(2) You do not live in California.	(2) $\sim Q$
Conclusion:	You do not live in Los Angeles.	$\therefore \sim P$

Modus tollens is a valid argument form since

$$[(P \rightarrow Q) \wedge (\sim Q)] \rightarrow \sim P$$

is a tautology. See the truth table of Figure 1.6.2. In examining the truth table for modus tollens we see that if $P \rightarrow Q$ and $\sim Q$ are true, then $\sim P$ is also true. As another example of the modus tollens type of argument,

FIGURE 1.6.2

P	Q	$P \rightarrow Q$	$\sim Q$	$(P \rightarrow Q) \wedge \sim Q$	$\sim P$	$[(P \rightarrow Q) \wedge \sim Q] \rightarrow \sim P$
T	T	T	F	F	F	T
T	F	F	T	F	F	T
F	T	T	F	F	T	T
F	F	T	T	T	T	T

Truth Table for Modus Tollens

consider the premises: (1) If you are a woman, then you are a human, and (2) You are not human. From these premises we may conclude: You are not a woman.

In addition to modus ponens and modus tollens, we will discuss a third type of valid argument appropriately referred to as the *chain rule* or the *rule of syllogism*. Aristotle taught that syllogisms were the main instruments for reaching logical conclusions. A syllogism has a particular form involving two premises and a conclusion. This form, which we will henceforth refer to as the **chain rule**, may be expressed symbolically as follows:

$$\text{Premises:} \qquad (1) \ \ P \to Q$$
$$(2) \ \ Q \to R$$

$$\text{Conclusion} \qquad \therefore \ \ P \to R$$

To prove the validity of this argument we show that

$$[(P \to Q) \land (Q \to R)] \to (P \to R)$$

is a tautology and hence always true. See Figure 1.6.3; note that we need eight rows in order to consider all possible truth values for the three statements P, Q, and R.

FIGURE 1.6.3

P	Q	R	$(P \to Q)$	$(Q \to R)$	$(P \to Q) \land (Q \to R)$	$P \to R$	$[(P \to Q) \land (Q \to R)] \to (P \to R)$
T	T	T	T	T	T	T	T
T	T	F	T	F	F	F	T
T	F	T	F	T	F	T	T
T	F	F	F	T	F	F	T
F	T	T	T	T	T	T	T
F	T	F	T	F	F	T	T
F	F	T	T	T	T	T	T
F	F	F	T	T	T	T	T

Truth Table for the Chain Rule
(Rule of Syllogism)

Example 2. A valid argument based on the chain rule is: If I live in Los Angeles, then I live in California; and if I live in California, then I live in the U.S.A. Therefore, if I live in Los Angeles, then I live in the U.S.A. The argument has the pattern:

Premises:	(1) If I live in Los Angeles, then I live in California.	(1) $P \rightarrow Q$
	(2) If I live in California, then I live in the U.S.A.	(2) $Q \rightarrow R$
Conclusion:	If I live in Los Angeles, then I live in the U.S.A.	$\therefore P \rightarrow R$

We have discussed three schemes of reasoning that constitute valid arguments: *modus ponens, modus tollens,* and the *chain rule.* These types of arguments are commonly used to assist in proving mathematical statements. There are, of course, other forms of valid arguments, but for now we will limit our discussion to these three.

EXERCISE 1.6

1. (a) Make a truth table for $[(P \rightarrow Q) \wedge P] \rightarrow Q$ (modus ponens). Compare your table with Figure 1.6.1.
   (b) Make a truth table for $[(P \rightarrow Q) \wedge \sim Q] \rightarrow \sim P$ (modus tollens). Compare your table with Figure 1.6.2.
   (c) Make a truth table for $[(P \rightarrow Q) \wedge (Q \rightarrow R)] \rightarrow (P \rightarrow R)$ (chain rule). Compare your table with Figure 1.6.3.
   (d) Is each of the implications in (a), (b), and (c) a tautology? Explain.

2. (a) Construct a truth table for $[(P \rightarrow Q) \wedge \sim P] \rightarrow \sim Q$.
   (b) Is $[(P \rightarrow Q) \wedge \sim P] \rightarrow \sim Q$ a tautology? Explain.
   (c) *Premises:* (1) If you practice a lot, you will win the tennis match, and (2) You do not practice a lot.
   Conclusion: You will not win the tennis match.
   Is this a valid argument? Explain.

3. Tell whether each of the following arguments is valid or invalid. If the argument is valid, indicate what type of argument is used.
   (a) If people aren't nice to me, I take my toys and go home. You weren't nice to me so I'm taking my toys and going home.
   (b) I always carry an umbrella if it rains. It is raining so I'll carry my umbrella.
   (c) If you own a Ford, you own a good car. John has a good car; therefore, he has a Ford.
   (d) If you own a Chevrolet, you own a good car. Mary doesn't own a good car; hence, she doesn't own a Chevrolet.
   (e) Men of distinction drink Double-Trouble. He drinks Double-Trouble; therefore, he is a man of distinction.
   (f) If you live in Miami, you live in Florida; and if you live in Florida, you live on a peninsula. Eugene lives in Miami; therefore, Eugene lives on a peninsula.

4. State the type of argument used in each of the following:

    (a) If a triangle has two congruent sides, it is an isosceles triangle. Triangle *ABC* has two congruent sides; therefore, it is an isosceles triangle.

    (b) If a triangle has two congruent sides, it is an isosceles triangle. Triangle *CDE* is not an isosceles triangle; therefore, it does not have two congruent sides.

    (c) If a triangle has two congruent sides, it is an isosceles triangle; and if a triangle is isosceles, its base angles are congruent. Triangle *JKL* has two congruent sides; therefore, triangle *JKL* has congruent base angles.

    (d) A whole number is an even number if and only if it has a factor of 2. The number 25 does not have a factor of 2; therefore, 25 is not an even number.

    (e) If a whole number is even, it has a factor of 2. If a number has a factor of 2, it is divisible by 2. The number 528 is even; hence, it is divisible by 2.

5. An interesting tautology is $(P \rightarrow Q) \vee (Q \rightarrow P)$.

    (a) Make a truth table to prove that $(P \rightarrow Q) \vee (Q \rightarrow P)$ is always true.

    (b) Must at least one of the following statements be true? Justify your answer.

        (1) If you read this book, you will go to Mars; or

        (2) If you go to Mars, you will read this book?

    (c) Which statement in part (b) is most likely to be true?

    (d) Could both statements in part (b) be true? Explain.

6. Using truth tables, test each of the following arguments and label them as valid or invalid:

    (a) *Given:* Florence will vote for Senator Killjoy or for Senator Luvall. However, she refuses to vote for Senator Killjoy.
    Conclusion: She must vote for Senator Luvall.

    (b) *Given:* If Sam fails his driver's test, he will be unable to drive. Sam passed his driver's test.
    Conclusion: Sam is able to drive.

    (c) *Given:* If Linda can understand logic, she can understand mathematics.
    Conclusion: Linda cannot understand logic or she can understand mathematics.

    (d) *Given:* If I drink coffee, it keeps me awake. I will not drink coffee.
    Conclusion: I will go to sleep.

1.7   Mathematical Reasoning and Deductive Proofs

When we attempt to prove certain results in mathematics, we generally use two types of reasoning—*inductive* and *deductive*. **Inductive reasoning** is a method of arriving at what seems to be a reasonable conclusion based on experimentation and observation of what happens in a number of similar situations. Even intuition resulting from a variety of past experiences may be of assistance in the inductive process. As an example of inductive reasoning, let us consider the set of odd numbers $\{1, 3, 5, 7, \ldots\}$ and observe that $1 = 1^2, (1 + 3) = 4 = 2^2, (1 + 3 + 5) = 9 = 3^2$, and $(1 + 3 + 5 + 7) = 16 = 4^2$. By induction it appears that $1 + 3 + 5 + 7 + 9$ will equal 5^2 and, in general, that $1 + 3 + 5 + \ldots$ to n addends will be equal to n^2. In this case our induction is correct, and we could *prove* that it is correct by deductive reasoning, a process that is discussed below.

Next, consider the equation $x = n^2 - n + 17$ where $n = 1, 2, 3$, and so on. Here we find that x has the values 17, 19, 23, 29, 37, 47, 59, 73, and 89, respectively, when n has the values 1, 2, 3, 4, 5, 6, 7, 8, and 9. Inductive reasoning may lead us to believe that x will be a prime number for every natural number n. (A prime number has exactly two distinct factors—itself and 1.) However, for $n = 17$ we get $x = 289 = 17 \times 17$. Hence 289 is not prime since it has three distinct factors, 1, 17, and 289, and our induction is false.

Since inductive reasoning can be unreliable, we depend upon other methods in attempting to prove statements in mathematics. But we do not discard a beautiful thing like inductive reasoning simply because it sometimes leads us astray. In fact, more often than not, induction tells us what might be true and, hence, worthy of an attempted proof. Frequently, having arrived at a tentative conclusion based on induction, we use **deductive reasoning** to *prove* (or disprove) our hypothesis or tentative conclusion.

In any deductive system where a set of statements is to be proved, it is necessary to start with certain **undefined terms**. Otherwise, the situation is similar to using an English dictionary without knowing the meaning of any English words. Of what value would it be to define *big* to mean *large* and *large* to mean *big,* if we knew the meaning of neither? Meaningful **definitions** can be formulated, however, by utilizing undefined terms and previously defined terms that may be available. In mathematics we might, for example, assume that everyone has at least an intuitive idea of what is meant by a set of points; then we could define a circle as a set of points having certain properties.

In a deductive system it is also necessary to assume certain basic concepts or statements which we accept as true without proof; in mathematics we call these assumptions **axioms** or **postulates**. We can then make *assertions* (other statements) which the basic assumptions seem to imply. When our assertions lead us in some logical way to a concluding statement, we say that this statement has been *proved*. If the proven statement is a general principle rather than an isolated fact, we may call it a **theorem**. The truth of the assertions in a proof may be justified by using any of the definitions, postulates, or previously proven theorems. Also, valid arguments such as modus ponens, modus tollens, and the chain rule may be explicitly stated in justifying assertions in a proof, or they may, as is frequently the case, be used in the general scheme of inference without explicit mention. In a lengthy proof the chain rule, as well as other methods of inference, may be used many times before the proof is completed.

Let us illustrate what has been said concerning the nature of proof by using a very simple direct proof to show that 14 is an even number. First, we define the whole numbers to be 0, 1, 2, 3, 4, and so on. We also define a whole number to be an even number if it has a factor of 2. Assuming that multiplication for the whole numbers has been defined, we are now prepared to prove that 14 is an even number.

Example 1(a).

To Prove: 14 is an even number.

Assertions	*Reasons*
1. 14 is a whole number.	1. By definition.
2. If a whole number has a factor of 2, then it is an even number.	2. By definition.
3. 14 has a factor of 2.	3. $14 = 2 \times 7$.
4. 14 is an even number.	4. Modus ponens applied to steps 2, 3, and 4.

The single isolated fact that 14 is an even number could hardly qualify for the distinction of being classified as a theorem. If, however, we were to prove a more general principle, such as the sum of any two odd numbers is an even number, or any even number greater than 4 may be expressed as the sum of two odd primes, then we might consider that we had proved a theorem.

Examining the proof in Example 1(a), we see that the first step merely establishes the fact that 14 is a whole number by the definition of whole numbers. (In a more difficult proof we would certainly omit this step rather than risk insulting the reader with the obvious.) The second assertion is of the form that says $P \rightarrow Q$ is true. The third assertion states that P is true. Hence we have the valid argument $[(P \rightarrow Q) \wedge P] \rightarrow Q$ (modus ponens), and we can state in step 4 that Q is true. This, of course, is what we wished to prove.

In actual practice we can, as shown below, shorten the above proof so that it has three steps rather than four.

Example 1(b).

To Prove: 14 is an even number.

Assertions	*Reasons*
1. 14 is a whole number.	1. By definition.
2. 14 has a factor of 2.	2. $14 = 2 \times 7$.

3. 14 is an even number. 3. If a whole number has a factor of 2, then it is an even number.

In the above proof modus ponens is still used but it is somewhat camouflaged. $P \rightarrow Q$ is the *third reason*, P is the *second assertion,* and Q is the *third assertion.* Hence we still used the fact that $[(P \rightarrow Q) \wedge P] \rightarrow Q$. In other words, since reason 3 and assertion 2 are true, assertion 3 is true.

In further modifying the proof that 14 is an even number, we could combine the assertions and reasons into ordinary sentences and present them using conventional exposition as follows:

Example 1(c).

To Prove: 14 is an even number.
By definition, 14 is a whole number. If a whole number has a factor of 2, then it is by definition an even number. Since $14 = 2 \times 7$, it has a factor of 2 and hence is an even number.

The ability to derive proofs in mathematics requires considerable study and experience. It is not something that you are unable to do one day and can suddenly master the following day. Proving statements or theorems in mathematics is probably somewhat of an art as well as a science; improvement comes with practice. There is no unique proof of a statement that is correct, while all others are wrong; *many* valid proofs may be written for a given proposition. A particular proof, however, may be easier to comprehend, more direct, or more convincing than some other proof, and hence more "elegant."

Just as there is no single set of correct steps for any given proof, there is no single acceptable format. Examples 1(a) and 1(b) of this section illustrate a commonly used format. First the proposition to be proved is stated. This is followed by a sequence of assertions along with reasons justifying the truth of each. The last assertion is, of course, the statement we hoped to prove. Example 1(c) of this section exemplifies another common method of organizing and presenting a proof.

1.8 Various Types of Mathematical Proofs

Mathematical proofs may be classified in various ways. The type of proof used in the previous section is usually referred to as a **direct proof**. The particular type of proof selected to prove a given statement will depend to some extent on the nature of that statement. If, for example, we wish to prove a proposition in which a limited number of cases is to be considered, we can test every case; this type of proof is called **proof by exhaustion**. (The name refers to exhausting all possible cases rather than to exhausting the person writing the proof!)

Example 1. Prove that every whole number greater than zero and less than ten that is a perfect square has an odd number of distinct factors.

Solution: Since $1 = 1^2$, $4 = 2^2$, and $9 = 3^2$, we see that 1, 4, and 9 are the only whole numbers between zero and ten that are perfect squares. The number 1 has only *one* distinct factor: 1. The number 4 has *three* distinct factors: 1, 2, and 4. The number 9 has *three* distinct factors: 1, 3, and 9. Hence, by the method of exhaustion, we see that every whole number between zero and ten that is a perfect square has an odd number of distinct factors.

A statement can be proved *not* to hold in general simply by showing one case in which it is false; this is known as a **proof by counterexample**.

Example 2. Is the statement "No mammals can fly" true?

Solution: No. Bats are mammals and bats can fly. This counterexample proves the statement to be false. Hence its negation must be true and we can conclude "Some mammals can fly."

Another type of proof used in mathematics is called the **indirect proof** or **proof by contradiction**. It is a useful type of proof but somewhat difficult to comprehend initially by a verbal description. The reasoning is as follows: We assume as true the *negation* of the statement we wish to prove and, by the implications of this assumption, arrive at a conclusion known to be false (a contradiction). If our implications are true, we conclude that we must have arrived at the contradiction because our assumption was false. Hence, the original statement rather than its negation must be true since exactly one of them is true.

Indirect proofs are not uncommon in everyday life. For example, in attempting to prove that his client is innocent of a crime, a lawyer might reason as follows:

> Assume that my client is *not* innocent. This implies that he must have been at the scene of the crime. But we have irrefutable evidence that he was not at the scene of the crime. This contradiction, of course, means that the assumption that my client is not innocent must be false. Hence, I declare his innocence.

Using the statements: (*P*) My client is innocent, and (*Q*) He was at the scene of the crime, we can present the above argument in outline form as follows:

Premises:	(1) If my client is not innocent, then he was at the scene of the crime.	(1) $\sim P \to Q$
	(2) My client was not at the scene of the crime.	(2) $\sim Q$
Conclusion:	My client is innocent (not not innocent).	$\sim(\sim P)$ or P

In examining the outline form of this indirect proof, we see that its use rests on the validity of the implication $[(\sim P \to Q) \wedge (\sim Q)] \to P$. This implication is a tautology; in fact, it is a modus tollens type of argument that can be verified by replacing P with $\sim P$ in the conventional modus tollens expression $[(P \to Q) \wedge (\sim Q)] \to \sim P$.

Although there are other methods used in mathematics, we have illustrated the nature of mathematical proof with some of the most common types. It is probably obvious by now that one could easily devote the entire text to logic and the nature of proof.

EXERCISE 1.8

1. (a) Using inductive reasoning, arrive at a tentative conclusion concerning whether or not the sum of any two even whole numbers will be odd or even.
   (b) Prove by deduction that the sum of any two even whole numbers is an even number. [*Hint:* An even number may be written in the form $2a$ where a is a whole number.]

2. (a) By induction, determine tentatively whether the product of two even whole numbers will be odd or even.
   (b) Prove by deduction that the product of any two even whole numbers is an even number.

3. (a) Prove that the sum of any two odd whole numbers is an even number. [*Hint:* An odd number may be written in the form $2a + 1$ where a is a whole number.]
   (b) Prove that the product of any two odd whole numbers is an odd number.

4. Prove by the method of exhaustion that no whole number from 1 to 9 inclusive has more than four distinct factors.

5. (a) Prove by a counterexample that it is false that every prime number is an odd number.
   (b) What related true statement can be made?

6. (a) Prove this statement to be false by a counterexample: No mammals live in the sea.
   (b) What related statement is true?

7. A baseball broke Mrs. Arnold's window at 10:15 a.m. She accused John of breaking the window. John was in school from 8:30 a.m. until noon. Write an indirect proof to show that John did not break the window.

REVIEW EXERCISES

1. Write the negation of each of the following statements:
   (a) Japanese citizens may not buy stock in American corporations.
   (b) Some Japanese citizens may buy stock in American corporations.
   (c) It is not the case that Japanese citizens may not buy stock in American corporations.

2. Make a truth table for $P \vee \sim Q$.

3. If B stands for "John plays baseball" and C stands for "John plays chess," make a truth table for "John does not play baseball and John plays chess."

4. State the converse, inverse, and contrapositive of the statement of implication "If you do not have assets of more than $20,000, then you are not a capitalist."

5. Let V stand for the statement "You have enough votes," and E stand for "You win the election." Make a truth table to show that the statement of implication "If you have enough votes, then you win the election" is equivalent to its contrapositive "If you do not win the election, then you do not have enough votes."

6. (a) Under what conditions will a statement of implication have a false converse?
   (b) Give an example of a statement of implication that has a false converse.

7. (a) Under what conditions will a statement of implication have a true converse?
   (b) Give an example of a statement of implication that has a true converse.

8. Tell whether each of the following arguments is valid or invalid. If the argument is valid, tell what type of argument is used.
   (a) If you speak well of others, they will speak well of you. Joan speaks well of others; therefore, others speak well of her.
   (b) If your car is to run, you must remember to put gas in the tank. Richard's car would not run; hence, he must have neglected to put gas in the tank.
   (c) If inflation is to slow down, we cannot have large deficit spending. However, we do have large deficit spending; therefore, inflation will not slow down.

(d) If you do not get enough sleep you will be tired, and if you are tired you cannot think well. Roger did not get enough sleep; hence, he cannot think well.

9. (a) Prove by a counterexample that the statement "All months have either 30 or 31 days" is false.

(b) What related statement is true?

10. Justify the modus tollens type of argument by showing that the truth table for $[(P \to Q) \wedge (\sim Q)] \to \sim P$ is a tautology (always true).

① A suff cond for liking math, is that you like this Book.

② All circle are Round

③ For me not to miss my breakfast, it is necessary that I do not get up late

④ for 7×5 to be = 35 , it is suff that 5+3= 7

⑤ if 3-×=1, then 2×5 = 10

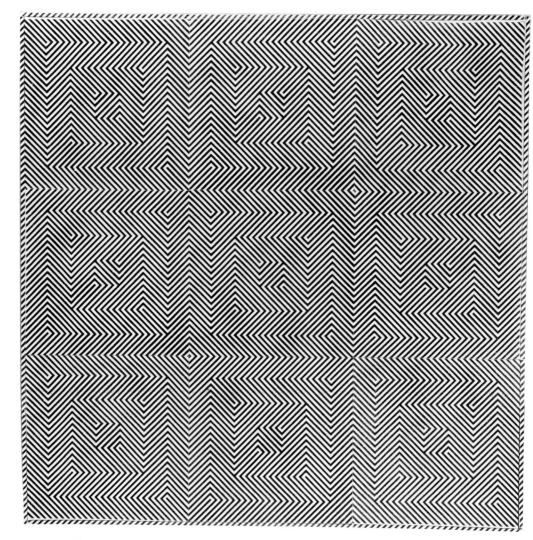

Reginald H. Neal, *Square of Three, Yellow and Black*
1964. Litho and paint on canvas. 32½ x 32½".
New Jersey State Museum

2 SETS, NUMBERS, AND NUMERALS

2.1 Introduction

Elementary mathematics probably began with attempts to answer such questions as, How many fish did you catch? How many women are in the cave? and How many days will you be gone? Basically these questions ask, How many elements are in a particular group or set of objects? The process we use in finding the number of elements in a set is, of course, **counting**. It seems appropriate to begin our discussion by developing a few elementary concepts concerning sets and determining how we count and what counting really is.

Although ideas about sets and simple numbers were undoubtedly developed prior to recorded history, the theory of sets was not established as a "respectable," important, and powerful branch of mathematics until after the publication of the works of Georg Cantor, a German mathematician who lived from 1845 to 1918.

Cantor's important works on set theory were created and published in the latter part of the nineteenth century (from about 1874 to 1895). Cantor once said, "The essence of mathematics resides in its freedom." However, the unconventional ideas resulting from his own freedom of expression met with strong initial resistance by the mathematics community. In particular, Cantor's former teacher Leopold Kronecker (1823–1891) aggressively opposed many of his ideas. Nevertheless, Cantor's writings opened up a new world for mathematicians to explore. Admittedly, there are paradoxes in some of Cantor's writings, but this in turn can lead to even more interesting studies. Much new mathematics has been developed in an attempt to eliminate the paradoxes of Cantor's original "naive" set theory.

Among those who were concerned about the paradoxes resulting from Cantor's set theory was the famous British philosopher and mathematician Bertrand Russell (1872–1970). An interesting and amusing paradox, for which Russell should be given credit, concerns the story of a certain village barber. The barber shaves those men and only those men in the village who do not shave themselves. Who then shaves the barber? He may not shave himself, for he shaves only those who do not shave themselves. No other person may shave the barber, for he himself shaves all of those who do not shave themselves. So who shaves the barber? This story, when translated into the language of sets, illustrates one of the problems with the original theory.

Many of the ideas in this chapter are related to concepts originated by Russell himself, and he is credited with having made significant contributions in defining the nature of number. A rather sophisticated treatise on the concept of a number is included in his book *Introduction to Mathematical Philosophy* published in 1919. This book was written by Russell while he was in a British prison during World War I. The governor of the prison, who was not a mathematician, was required to read all of Russell's writings for possible seditious content; hence, he was probably happy when Russell was released. (Russell was imprisoned for his outspoken pacifistic views. Holding essentially the same basic philosophy over 30 years later, he received the 1950 Nobel Prize for literature for his writings "as a defender of humanity and freedom of thought.") Undoubtedly, Russell was one of the most influential forces in modern science and he ranks among the great mathematicians of history. Alfred North Whitehead (1861–1947), another outstanding mathematician and philosopher of recent times, collaborated with Russell over a period of ten years and together they produced the publication *Principia Mathematica;* this work is believed by many to be the greatest single contribution to logic since that of Aristotle.

In the same sense that the atom is a "building block" for the physicist and the cell is a building block for the biologist, concepts of sets and numbers are building blocks for the mathematician. By studying the nature of the atom, the physicist searches for new sources of energy; by studying the nature of the cell, the biologist searches for a cure to cancer; and by studying set theory and the nature of number, the mathematician develops new concepts in both pure and applied mathematics.

2.2 The Concept of a Set

The meaning of the word set in mathematics is essentially the same as in everyday language. We all know what is meant when we speak of a set of books, a set of dishes, or a salt and pepper set. In mathematics the members of a set are also called elements of the set. The elements of a set may be physical objects such as automobiles, books, and molecules of water, or they may be abstract elements such as letters of the alphabet, numbers, points, lines, planes, axioms, or ideas. Mathematics places no limitation on the nature of the elements that may be considered as members of a set.

Although we have illustrated what is meant by a set of elements, the terms "set" and "element" will not be formally defined. As we mentioned in Section 1.7, it is necessary in any system of logic to have some undefined terms, and it seems reasonable to assume that the reader has acquired through experience a concept of what is meant by a set of elements. Therefore, we shall simply seek to characterize various sets by indicating the elements that belong to them.

2.3 Notation for Sets

One way to indicate a set is to list the names of all the elements in roster form and enclose them in a pair of braces. For example, the set whose elements are 1, 3, 5, and 7 may be indicated as $\{1, 3, 5, 7\}$. Note that the names of the elements are separated by commas. The order in which the elements of a set are listed makes no difference except possibly as a matter of convenience. In the above example the set could have been indicated as $\{3, 5, 7, 1\}$ or $\{7, 1, 3, 5\}$ or in any other way obtained by reordering the elements. (Can you determine how many reorderings are possible?)

If a set has many elements, we sometimes name enough of them to make the pattern clear and indicate the missing elements with three dots. For example, the set of letters in our alphabet may be indicated as $\{a, b, c, \ldots, z\}$. The set of odd whole numbers, which has infinitely many elements, is $\{1, 3, 5, \ldots\}$. The set of integers is $\{\ldots, -2, -1, 0, 1, 2, \ldots\}$.

If we do not wish to list any elements of a set, we can let a capital letter such as A, B, or C represent the set. Hence $A = \{1, 3, 5, 7\}$ is read, "A is equal to the set whose elements are 1, 3, 5, and 7." Capital letters are used to denote sets; lower case letters are frequently used to represent elements of a set. If we wish simply to identify a set of any four elements, we might use $A = \{a, b, c, d\}$. If we then wish to identify some other set containing three elements, we can let $B = \{e, f, g\}$.

Another important symbol in dealing with sets is $\in$, which is read "is an element of." For example, $3 \in A$ is read, "3 is an element of set A" or "3 is a member of set A." Ordinarily a slanted bar, /, through a symbol negates it. Thus $9 \notin A$ is read, "9 is not an element of set A." Similarly, $3 \neq 5$ is read, "3 is not equal to 5."

Sometimes it is convenient or even necessary to describe the elements of a set rather than to list them in roster form, particularly if there are a great many elements. For this purpose we use what is commonly known as **set-builder notation**. The expression

$$A = \{x \mid x \text{ has a certain property}\}$$

is read, "A is equal to the set of all elements x such that x has a certain property." The vertical bar is the only new symbol we have used and in this context it is read "such that." The set $\{1, 3, 5, 7\}$ could be indicated by writing $\{x \mid x \text{ is a positive odd number less than } 8\}$; this is read, "the set of all elements x such that x is a positive odd number less than 8." The set of all odd integers can be indicated by simply writing $\{x \mid x \text{ is an odd integer}\}$; this is read, "the set of all elements x such that x is an odd integer." Observe that the restrictions placed on the elements of the set follow the vertical bar.

Let us emphasize that there is no unique, correct method for indicating a set using set-builder notation; several different and perhaps equally useful methods for identifying the elements of any given set may be formulated. However, the defining property of a set in mathematics should be carefully and clearly stated so that it is possible to tell whether or not any given element is a member of the set; such a set is said to be **well-defined**. (Note that we are not attempting to define the general notion of a set but rather to characterize a particular set.) For example, it is easy to determine membership in the set whose elements are the first ten letters of the alphabet, but little agreement could be expected concerning membership in the set described as the set of all good-looking girls.

EXERCISE 2.3

1. State very briefly your reaction to the statement, "New ideas in mathematics, science, and medicine are readily accepted, while new ideas in sociology, religion, and philosophy are strongly resisted."

2. Which of the following sets are well-defined?
   (a) Great movies of the twentieth century.
   (b) Positive integers greater than 8 but less than 20.

(c) The first five letters of our alphabet.

(d) The ten best dressed women in the world.

(e) Cities with a lot of smog.

3. Use the roster method to express each of the following sets in as many ways as possible.

(a) $\{5, 4, 3\}$ (b) $\{a, b\}$

(c) $\{9\}$ (d) $\{a, b, c\}$

4. Express in words the meaning of the following:

(a) $A = \{8, 9, 10\}$

(b) $8 \in A$

(c) $\{x \mid x = \text{a whole number}\}$

(d) $\{x \mid x \text{ is an integer greater than 5}\}$

(e) $B = \{x \mid x \in A \text{ and } x \neq 8\}$

(f) $K = \{b, c, 5\}$

(g) $5 \notin K$

5. Use mathematical symbols to indicate the following:

(a) C is equal to the set whose elements are 5, 10, and 15.

(b) 5 is a member of set C.

(c) 7 is not an element of set C.

(d) The set of all elements x such that x is an even integer.

(e) S is equal to the set of all elements y such that y is an odd number less than 7.

6. Use set-builder notation to indicate each of the following sets:

(a) $\{\ldots, -2, -1, 0, 1, 2, 3, \ldots\}$

(b) $\{1, 3, 5, 7, 9\}$

(c) $\{4, 6, 8, 10, 12, \ldots\}$

(d) $\{a, b, c, d, e, f, g\}$

(e) $\{s, t, u, v, w, x, y, z\}$

(f) $\{\ldots, -2, -1, 0, 1, 2\}$

2.4 Equal Sets and the Empty Set

Sometimes sets that are defined differently may actually have the same elements. In this event the sets are said to be *equal*.

DEFINITION 2.4.1. If every element of set A is an element of set B and if every element of set B is an element of set A, then set A is **equal** to set B.

We denote that set A is equal to set B by writing $A = B$. At first it may seem a waste of time to call one set A and another set B if they are indeed the same set. However, in working with two sets we may not immediately realize that they are the same set. Note that by definition equal sets contain exactly the same elements and that A and B are two names for the same set.

Usually it serves no useful purpose to name the same element in a set more than once since, for example, $\{a, a, b\} = \{a, b\}$. However, if a set happens to contain two or more elements that are not distinguishable from each other, this can be made clear by giving them related but different names. For example, in printing the word *alfalfa* in this text, we need three elements of a, two of l, and two of f. We could identify this set by using subscripts and writing $\{a_1, a_2, a_3, l_1, l_2, f_1, f_2\}$ which is read, "the set whose elements are a sub-one, a sub-two," and so on.

Example 1.   Suppose set A contains every student in a class who owns an automobile, and set D contains every student who has a driver's license. If every student who owns an automobile has a driver's license, and if every student with a driver's license owns an automobile, then set A is equal to set D.

Example 2.   The set $\{a, 4, 8, q\}$ is equal to the set $\{4, 8, q, a\}$ since they have the same members, and we can write $\{a, 4, 8, q\} = \{4, 8, q, a\}$.

A set may be defined but contain no elements; for example, consider the set of men 200 feet tall or the set defined by $\{x \mid x$ is a whole number between 7 and 8$\}$. Many other examples can be given. Such a set is called *the empty set*.

DEFINITION 2.4.2.   The set containing no elements is called the empty set.

The empty set is denoted by the symbol $\emptyset$ which may simply be read as "the empty set." The empty set may also be denoted by a pair of empty braces $\{\ \}$, but the symbol $\emptyset$ is usually preferred and is easier to write. Note that the empty set is *not* designated by $\{\emptyset\}$; this notation would represent a set containing one element $\emptyset$ rather than the set with no elements. Similarly, the set $\{0\}$ contains the element zero and hence is *not* the empty set. Since, by Definition 2.4.1, any two empty sets would have to be equal, there is only one empty set. Hence the expression "*the* empty set" is used rather than "*an* empty set." The empty set may also be called the null set or the void set.

2.5   Subsets and Proper Subsets

Sometimes we find that every member of one set is also a member of some other set. If such a relation exists, the one set is said to be a *subset* of the other. For example, the set of insects is a subset of the set of all animals in the world.

✕ DEFINITION 2.5.1.   If every element of set A is also an element of set B, then set A is a subset of set B.

We indicate that A is a subset of B by writing $A \subseteq B$. This is read, "A is a subset of B."

Example 1.   If $A = \{a, b\}$ and $B = \{a, b, c\}$, then $A \subseteq B$ since every element in A is also an element in B. The fact that A is a subset of B may also be expressed by writing $\{a, b\} \subseteq \{a, b, c\}$.

Example 2.   If $C = \{a, d\}$ and $A = \{a, b, c\}$, then C is not a subset of A since $d \in C$ but $d \notin A$. We can then write $C \nsubseteq A$ which is read, "C is not a subset of A."

It follows from Definition 2.5.1 that every set is a subset of itself. Furthermore, since the empty set contains no elements, it is necessarily true that there are *no* elements in the empty set that are *not* members of every other set; hence, the empty set is a subset of every other set, including itself. Problem 4 of Exercise 2.5 further clarifies this situation.

Frequently we wish to refer to part of a set but not the entire set. In this case we are dealing with a set known as a *proper subset* of the original set.

DEFINITION 2.5.2.   If every element of set A is an element of set B, but not every element of set B is an element of set A, then set A is a proper subset of set B.

The fact that A is a proper subset of B is denoted by writing $A \subset B$. This is read, "A is a proper subset of B." Note that by Definition 2.5.2 every subset of any given set B, including the empty set, is a proper subset of B except B itself.

Example 3.   If $S = \{a, b, c\}$, the subsets of S are $\{a, b, c\}$, $\{a, b\}$, $\{a, c\}$, $\{b, c\}$, $\{a\}$, $\{b\}$, $\{c\}$, and $\varnothing$. By definition, all these are proper subsets of S except S itself or $\{a, b, c\}$.

Now that we have defined equal sets, subsets, and proper subsets, it is interesting to consider the relationship that exists between the symbol $\subseteq$ and the symbols $\subset$ and $=$. The symbol $\subseteq$, used to indicate that one set is a subset of another, is a combination of $\subset$, the symbol for a proper subset, and $=$, the symbol for equals. Since a subset is either a proper subset of a given set or equal to the given set, this symbolism is quite appropriate.

EXERCISE 2.5

1. Define each of the following terms:
   (a) equal sets   (b) the empty set
   (c) subset   (d) proper subset

2. Express in words the meaning of the following:
   (a) $A \subset B$ (b) $A \subseteq B$
   (c) $\{a, b\} \subseteq \{a, b\}$ (d) $\{a, b\} \not\subset \{a, b\}$
   (e) $\varnothing \subseteq \{a, b\}$ (f) $\varnothing \subset \varnothing$
   (g) $\varnothing \not\subset \varnothing$

3. Which of the following sets are equal?
   $A = \{1, 3, 5, 7\}$
   $B = \{2, 4, 6, 8\}$
   $C = \{x \mid x$ is an odd whole number and x is less than $9\}$
   $D = \{5, 3, 7, 1\}$

4. Given $A = \{a, b, c, d\}$ and the empty set $\varnothing$:
   (a) Is every element in $\varnothing$ in A? (If your answer is "no," name an element in $\varnothing$ that is not in A.)
   (b) Is every element in A in $\varnothing$?
   (c) Is $\varnothing$ a proper subset of A by Definition 2.5.2?

5. Explain the difference between $\varnothing$ and $\{\varnothing\}$.

6. Which of the following sets are subsets of $\{a, b, 7, \triangle\}$?
   $A = \{a, c\}$ $B = \{7\}$
   $C = \{5, 7\}$ $D = \{a, b, 7, \triangle\}$
   $E = \varnothing$ $F = \{\varnothing\}$
   $G = \{\triangle, a, 7\}$ $H = \{\triangle, b, 7, a\}$
   $I = \{a, b, \varnothing\}$ $J = \{\ \}$

7. Which of the sets in Problem 6 are proper subsets of the given set?

8. State the number of elements in each of the following sets.
   (a) $\{7, a, \triangle, \varnothing\}$ (b) $\{a, \triangle, \varnothing\}$
   (c) $\{\triangle, \varnothing\}$ (d) $\{\varnothing\}$
   (e) $\varnothing$

9. Which of the following sets are equal?
   (a) $\{\ \}$ (b) $\{\varnothing\}$ (c) $\{0\}$
   (d) $\varnothing$ (e) $\{0, \varnothing\}$

2.6   Equivalent Sets and the Concept of Number

To discuss the number concept we first must understand what is meant by a *one-to-one correspondence* between the elements of two sets. In everyday language we might say that the elements of two sets are in a one-to-one correspondence if they are paired in such a way that every element in each of the sets has exactly one mate in the other set. There are many examples of sets whose elements can be placed in a one-to-one correspondence. A shoe salesman should find a one-to-one correspondence between the elements of the set of left shoes and the elements of the set of right shoes in his store. In a theater there should be a one-to-one correspondence between patrons and tickets sold. A one-to-one correspondence should exist between husbands and wives in a monogamous society.

DEFINITION 2.6.1.   The elements of two sets A and B are said to be in a **one-to-one correspondence** if each member of A is paired with exactly one member of B and if each member of B is paired with exactly one member of A. When an element of A is paired with an element of B, that element of B is automatically paired with the element of A.

In Figure 2.6.1, (a) and (b) are one-to-one correspondences since they each satisfy Definition 2.6.1. Note that arrows go from elements of set A to elements of set B and also from elements of set B to elements of set

FIGURE 2.6.1

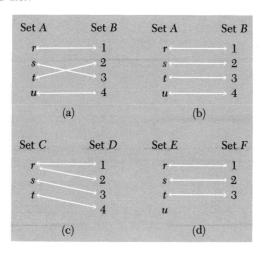

A. The double-headed arrows indicate that whenever an element of set A is paired with an element of set B, that element of set B is automatically paired with the element of A. Part (c) of Figure 2.6.1 is not a one-to-one correspondence since two elements of set D are paired with r, and part (d) is not a one-to-one correspondence since u is not paired with any element of set F.

The elements of some sets can be paired in many different ways to produce a one-to-one correspondence. Those of sets A and B in Figure 2.6.1 are paired in two different ways; but they could be paired in 24 different ways, as we will discover in Chapter 8.

DEFINITION 2.6.2.   If the elements of two sets A and B can be placed in a one-to-one correspondence, the sets are said to be equivalent.

When two sets are equivalent, we have one *set* related to another *set*. To be sure, it is only when the *elements* of two sets can be placed in a one-to-one correspondence that the sets are related—but it is the *sets* (not the elements) that are equivalent.

All sets that are equivalent are said to belong to the same equivalence class. For example, $\{a, b, c\}$, $\{4, 7, 9\}$, $\{\triangle, 2, f\}$, and $\{cat, horse, ox\}$ are equivalent sets and belong to the same equivalence class. The property that these equivalent sets have in common is called their *cardinal number.* We happen to give the cardinal number of these sets the name "three." Similarly, the sets $\{b, a\}$, $\{6, 4\}$, and $\{\triangle, e\}$ are members of the same equivalence class and we say that each of these sets has the cardinal number "two."

DEFINITION 2.6.3.   The property that is possessed by all members of a class of equivalent sets is the cardinal number of each of the sets.

Cardinal # = Always answers the question, how many?

The important result of this definition is that any two sets that are equivalent have the same cardinal number. A number is an abstraction and should not be confused with a *name* for the number. A name or symbol for a number is called a numeral. Both III and 3 are numerals for a certain number. Many different numerals can be invented for any given number.

To designate the cardinal number of a set A, we use the symbol $n(A)$, which is read, "the number of elements in set A." For example, if $A = \{a, b, c\}$, we can write $n(\{a, b, c\}) = 3$ or $n(A) = 3$. In saying that $n(\{a, b, c\}) = 3$, we have assumed that the reader can count. But what is counting? This question will be answered in the next section.

Ordinal — always answers the question, what order.
Identification — identifies something.

2.7   Standard Sets, Natural Numbers, and Whole Numbers

The elements of sets such as $\{a\}$, $\{b\}$, $\{\triangle\}$, and $\{k\}$ can be placed in a one-to-one correspondence; the sets are, therefore, equivalent to each other and have the same cardinal number. We use the symbol 1 (the numeral 1) to name the cardinal number of each of these sets as well as the cardinal number of each of the other members of the equivalence class to which these sets belong. We can also identify a set containing only the element 1 denoted by {1}. We call this particular set the standard set for the class of sets that are equivalent to it. Similarly, $\{a, b\}$, $\{\triangle, k\}$, and $\{\triangle, t\}$ are members of another equivalence class of sets, and we name the cardinal number of each of these sets 2. By using all the elements of the previous standard set $\{1\}$ (there is a single element in this case) together with the element 2, we can generate a standard set $\{1, 2\}$ for this class. In the same way we can form the standard sets $\{1, 2, 3\}$, $\{1, 2, 3, 4\}$, ..., $\{1, 2, 3, 4, 5, \ldots, n\}$. Each time we generate another standard set, we use all the elements of the previously formed standard set together with one more element. We see that sets formed in this way are *ordered* and that the cardinal number of each set is the last-named element of the set: $n(\{1\}) = 1$, $n(\{1, 2\}) = 2$, $n(\{1, 2, 3\}) = 3$, ..., and $n(\{1, 2, 3, \ldots, k\}) = k$.

If the set $\{1, 2, 3, 4, 5, 6, \ldots\}$ is continued without termination, as indicated by the dots, it is said to be an *infinite* set and to contain infinitely many elements. This set is called the set of natural numbers or counting numbers and is designated as $N = \{1, 2, 3, \ldots\}$.

The standard sets used to represent sets belonging to the same equivalence class are initial subsets (occur at the beginning) of the natural numbers. This means that the cardinal number of any finite set, except the empty set, may be found by placing the elements of the set in a one-to-one correspondence with the elements of an initial subset of the natural numbers. The last named natural number in the correspondence will name the cardinal number of the set. This process is called counting. When a child counts "one, two, three" on his fingers, he is putting his fingers in a one-to-one correspondence with the elements of an initial subset of the natural (or counting) numbers.

Example 1. Find the cardinal number of $\{a, b, \varnothing, k, \triangle\}$.

Solution: First establish a one-to-one correspondence between elements of the given set and those of an initial subset of the natural numbers:

$$\{a, b, \varnothing, k, \triangle\}$$
$$\updownarrow \;\; \updownarrow \;\; \updownarrow \;\; \updownarrow \;\; \updownarrow$$
$$\{1, 2, \;\; 3, \;\; 4, \;\; 5, 6, 7, \ldots\}$$

Since the elements of the given set can be placed in a one-to-one correspondence with the elements of the standard set $\{1, 2, 3, 4, 5\}$, and since the last named element of this set is 5, we know that $n(\{a, b, \varnothing, k, \triangle\}) = n(\{1, 2, 3, 4, 5\}) = 5$.

It is important to emphasize that the standard sets, $\{1\}$, $\{1, 2\}$, $\{1, 2, 3\}$, and so on, are ordered sets since they are initial subsets of the set of natural numbers $\{1, 2, 3, \ldots\}$, which is itself an ordered set. If the standard sets were not ordered, we could not name the cardinal number of a given set by its last-named element, since this would result in ambiguity.

It was quite an achievement for man to invent names for each of the natural numbers whereby one can continue the sequence indefinitely without a lot of memorization. The pattern is so simple and effective that nearly everyone can identify the number that comes after any given number, such as 284, or 72,394, or 285,623,435. (Can you think of any other infinite set in which you can easily name each of the elements in succession?) Contributing to the simplicity of the system is the fact that in naming all the natural numbers, a set of only ten basic elements is used: $\{0, 1, 2, 3, 4, 5, 6, 7, 8, 9\}$.

Until now the cardinal number of the empty set has been avoided. We avoid it no longer and define the cardinal number of the empty set to be 0 (zero) and write $n(\varnothing) = 0$. If we take the set of natural numbers, $N = \{1, 2, 3, \ldots\}$, and include zero with it, the resulting set is the **whole numbers** $W = \{0, 1, 2, 3, \ldots\}$. (The cardinal number of infinite sets such as N and W will be discussed in Chapter 5.)

EXERCISE 2.7

1. Define *one-to-one correspondence*. ← *need same number of elements in each set.*

2. Define *equivalent sets*.

3. Explain the difference between equivalent sets and equal sets. Refer to Section 2.4 if necessary.

4. (a) Are equal sets necessarily equivalent? Yes
   (b) Are equivalent sets necessarily equal? NO

5. Given:
   $A = \{a, b, c\}$
   $B = \{c, a, b\}$
   $C = \{1, 2, 3\}$
   $D = \{b, a, c\}$
   $E = \{4, 5, 6\}$
   (a) Which sets are equal? A BD
   (b) Which sets are equivalent? All

6. (a) In how many ways can the elements of $\{a, b, c\}$ be placed in a one-to-one correspondence with the elements of $\{1, 2, 3\}$?
   (b) Show all possible one-to-one correspondences.

7. (a) Can the elements of $\{a, b, c, d, e\}$ be placed in a one-to-one correspondence with themselves?
   (b) Is every set equivalent to itself?
   (c) Is every set equal to itself?

8. Define the cardinal number of a set.

9. Find the number of elements in $\{a, \triangle, 4, \varnothing, 8\}$ by placing the elements of the set in a one-to-one correspondence with the elements of an initial subset of the natural (counting) numbers.

10. Identify the standard set that is equivalent to each of the following sets and give the name of the cardinal number of each set.
    (a) $\{2, 4, 6\}$ (b) $\{a, b\}$
    (c) $\{3, 6, 9, 12, 15\}$ (d) $\{g\}$
    (e) $\{\varnothing, A, 7, 9\}$

11. Explain what is meant by "counting" the elements of a set.

12. If every chair in a classroom has a person sitting in it and every person in the room is sitting in a chair:
    (a) Are the set of chairs and the set of people equivalent sets? Explain.
    (b) Are the set of chairs and the set of people equal sets? Explain.
    (c) Do the set of chairs and the set of people have the same cardinal number? Explain.
    (d) How would you find the cardinal number of each set?

2.8 Early Numeral Systems

In Section 2.6 we said that the name or symbol for a number is called a *numeral*. A plan by which symbols of a given set are used separately or in combination to form numerals is called a **system of numeration**. Although many systems of numeration have been invented, we shall examine only a few. After studying this section, the reader may be inspired to try to originate his own system of numeration.

One of the earliest systems of numeration (probably the earliest) was the *tally system*. This system, which was only one easy step beyond making a one-to-one correspondence between such things as pebbles and sheep, consisted merely of making strokes in the sand or in clay in a one-to-one correspondence with the members of a flock of sheep, a group of warriors, a herd of cattle, or some other set. Even today we sometimes use tally marks to record such simple things as the score in a game or the number of persons enrolling in a course.

In Egyptian and Grecian lands primitive numerals appeared as $|$, $||$, $|||$, and so forth, while in the Far East the tally system appeared as —, $=$, $\equiv$, and so on. Today we frequently group at five as a matter of convenience and write the numeral twelve, for example, as ⊥⊦⊦⊦ ⊥⊦⊦⊦ $||$.

Originally, the word "tally" referred to a stick with cross notches on it indicating the amount of a debt. The stick would be split lengthwise with one half going to the debtor and the other half to the creditor. The two halves could then be matched at any future time to verify the amount of the debt.

One of the earliest systems of numeration on record is the *Egyptian system,* which dates back as far as 3000 B.C. Early hieroglyphic (picture) symbols used in the Egyptian numeration system are shown in the chart of Figure 2.8.1 and compared with the equivalent Hindu-Arabic numerals that we use. The figure shows that the Egyptians, who lived along the Nile River, represented numbers with pictures of familiar objects including the tadpole and the lotus flower.

FIGURE 2.8.1

Hindu-Arabic Numeral	Egyptian Numeral	Description of Symbol
1	I	Stroke
10	∩	Heel bone
100	℮	Scroll
1,000	⚶	Lotus flower
10,000	⌒	Bent finger
100,000	⚮	Tadpole
1,000,000	⚸	Astonished man

In studying Figure 2.8.1 you will see that a single symbol is used for each number that is a power of ten: 1, 10, 100, 1000, and so forth. Numerals for numbers other than powers of ten are formed simply by repeating a symbol and using the additive principle. For example, 32 may be written ∩∩∩II. Numerals from one to twenty are shown in Figure 2.8.2, and other selected numerals are shown in Figure 2.8.3. Since only the additive principle is involved in writing Egyptian numerals, the arrangement of the symbols for any given number is simply a matter of convenience. For example, fifteen may be written as shown in Figure 2.8.2 or as ∩IIIIII, or IIIIII∩, or II∩III, and so on.

The Egyptian system of numeration had two distinct disadvantages. It contained no numeral for zero and did not use the concept of *place value.* By place value we mean the modification of a symbol's value according to its location or place in a numeral. In our system of numeration, for example, $52 \neq 25$ even though the same symbols, 2 and 5, are used in each numeral. However, since place value was not a feature of the Egyptian system, the order in which symbols appeared in a numeral was immaterial.

Another early system of numeration was that of the Babylonians. The early *Babylonian system of numeration* dates back to about the same time

FIGURE 2.8.2

one	two	three	four	five	six
I	II	III	IIII	III II	III III

seven	eight	nine	ten	eleven	twelve
IIII III	IIII IIII	III III III	∩	∩I	∩II

thirteen	fourteen	fifteen	sixteen
∩III	∩IIII	III ∩II	III ∩III

seventeen	eighteen	nineteen	twenty
IIII ∩III	IIII ∩IIII	III III ∩III	∩∩

as the Egyptian system (3000 B.C.). Babylonian writings have been preserved in clay that was baked in the sun or in kilns. Numerals in the Babylonian system were formed with cuneiform (wedge-shaped) symbols. Like the Egyptian system, the Babylonian system had an additive property. However, only two symbols were used: ▼ for one and ◀ for ten. Certain selected numerals are shown in Figure 2.8.4.

The Babylonian numeral system also had a place value property which led to its designation as a *sexagesimal* system. The place of the symbol in the numeral determined the power of 60 by which its value was multi-

FIGURE 2.8.3

Hindu-Arabic Numerals	Egyptian Numerals
72	∩∩∩∩ ∩∩∩ II
93	∩∩∩ ∩∩∩ III ∩∩∩
262,323	∞ ∞ ⌠⌠⌠ ⌠⌠⌠ ⌡⌡ ℮℮℮ ∩∩ III
1,020,052	⚡ ⌠⌠ ∩∩∩ ∩∩ II

FIGURE 2.8.4

Did have place value

Hindu-Arabic Numeral	2	5	8	14	24
Babylonian Numeral	▼▼	▼▼▼ ▼▼	▼▼▼ ▼▼▼ ▼▼	⟨▼▼▼ ▼	⟨⟨▼▼▼ ▼

plied. Beginning at the right, the various groups of symbols would be multiplied by 60^0, 60^1, 60^2, 60^3, and so on. (Incidentally, we shall show later that $60^0 = 1$.) The numeral ▼▼ ⟨▼ ▼▼▼, for example, would have the value $(2 \cdot 60^2) + (11 \cdot 60^1) + (3 \cdot 1)$, which we would write as $7200 + 660 + 3 = 7863$. Although we have spaced the symbols appropriately in the Babylonian numeral above, this was frequently not done by the Babylonians, and only the context of the writing could be used to tell which symbols belonged in the units place, the 60's place, the $(60 \cdot 60)$'s place, and so on. As a result, several interpretations were frequently possible.

Even though the Babylonian numeration system was often ambiguous, had no symbol for zero, and was awkward to use in many ways, it was a great step forward because of its place-value characteristic. Clay tablets of the Babylonians with their cuneiform symbols are displayed in some of our museums and are worth a visit. We can probably thank the stable nature of the clay tablets and the dry climate of Babylonia for their preservation. Remnants of the early Babylonian numeral system may be found in contemporary cultures. For example, there are $6 \cdot 60$ or 360 degrees in a circle, 60 seconds in a minute, and 60 minutes in an hour.

Other than the Hindu-Arabic system of notation, the *Roman system* is probably the most familiar. It uses letters as numerals. Basic symbols are given below with the corresponding Hindu-Arabic symbol:

always a power of

I	V	X	L	C	D	M
1	5	10	50	100	500	1000

The rules for writing Roman numerals are somewhat complex. Essentially the Roman system is an *additive* system with *subtractive* and *multiplicative* features. If the symbols decrease in value from left to right, their values are to be added; however, if a symbol has a smaller value than the symbol to its right, subtraction is indicated. For example, XI = $10 + 1 = 11$, but IX = $10 - 1 = 9$. In general, not more than four identical symbols are used in succession since a new symbol can replace them. For example, IIII is not used for 5 since V can be used instead. In fact, usually not more than three identical symbols are used in succession since the subtractive property is available: 4 is written as IV rather than IIII, and 90 is XC rather than LXXXX.

In applying the subtractive feature of the Roman numeral system, we may write only symbols representing numbers that are powers of ten (I, X, and C) to the left of symbols representing larger numbers. Furthermore, a symbol can only be written to the left of symbols for the next two larger numbers having distinct symbols. Hence only subtraction combinations of IV, IX, XL, XC, CD, and CM are found in Roman numerals, while combinations such as IL and XD are not used. Note that there are never more than two symbols involved in a subtraction combination. Without definite rules for the subtractive feature, ambiguous situations would result. For example, what number would be represented by IVX, 6 or 4? The answer, of course, would depend upon whether we subtract I from V first or V from X first.

As an alternative to making new symbols for very large numbers, the multiplicative property is used. One bar over a numeral multiplies its value by 1000, and two bars over a numeral multiplies its value by $1000 \cdot 1000$ or 1,000,000. For example, $\bar{V} = 5 \cdot 1000 = 5000$; $\overline{\overline{IX}} = 9 \cdot 1,000,000 = 9,000,000$; and $\overline{CCCDCII} = 300 \cdot 1000 + 500 + 100 + 1 + 1 = 300,602$.

Because of the influence of the Roman Empire and the fact that the system was as good or better than most other systems that had been developed, the Roman numeral system held a strong position for nearly 2000 years in commerce and in scientific and theological literature. The Hindu-Arabic system finally replaced the Roman system because of its overwhelming superiority, but it was a bitter battle that was not easily won. The Hindu-Arabic system has been in general use only about 400 years, which historically is a very short time.

Occasionally Roman numerals are still used for such things as chapter headings in books and corner stones of buildings because of their esthetic appeal and decorative effect. Apparently, they are also used to camouflage the dates on which movies are released. (Take note the next time you see a movie in a theater or on television.)

EXERCISE 2.8

1. Do you believe it is likely that the tally system of numeration was invented independently by various people? Explain.

2. Write an Egyptian numeral for each of the following:
   (a) 43   (b) 3628   (c) 2,300,562

3. How does the number represented by the Egyptian numeral ℮ ⸑ ‖ ∞ compare with the number represented by ∞ ⸑ ℮ ‖?

4. Using our system of numeration, write the equivalent of each of the following Egyptian numerals:

(a) ⲅⲅ ∩∩∩ ∩∩∩ |||

(b) 𓆏 ⲇⲇ ∩∩∩∩∩ ||| ∩∩∩∩ |||

(c) ⲅⲅⲅ ⲇⲇ ⲉⲉⲉ ∩∩||||| ∩∩|||

(d) ⲅⲅⲅ ⲉⲉⲉ |||

5. Referring to Problem 4, use Egyptian numerals to find the sum of (a) and (b).

6. Write Babylonian numerals for the following:
   (a) 4   (b) 32   (c) 612

7. Which basic property of the Babylonian numeral system is not found in the Roman system?

8. Write decimal numerals for the following Roman numerals:
   (a) XXIV 24 (b) CCXXIV
   (c) DCCLXXXII 782 (d) MMLVIII

9. Write Roman numerals for each of the following:
   (a) 1215 $MCCXV$ (b) 1492
   (c) 1066 $MLXVI$ (d) 3,000,000

10. Write the three Roman numerals immediately following DCXLVIII.

11. Referring to Problem 8, use the Roman numeral system to find the sum of (a) and (b).

2.9 Exponents

We shall find it helpful for our discussion of the Hindu-Arabic numeral system to review a few concepts about the use of *exponents*. You will probably recall that we can write $5 \cdot 5 \cdot 5$ as 5^3, and $6 \cdot 6 \cdot 6 \cdot 6$ as 6^4. In general, $a^n = a \cdot a \cdot a \cdots$ to n factors of a. The a is called the base and the superscript n is called the exponent of a or the power to which a is raised.

Example 1. $2^5 = 2 \cdot 2 \cdot 2 \cdot 2 \cdot 2 = 32$.

Example 2. $10^3 = 10 \cdot 10 \cdot 10 = 1000$; $10^2 = 10 \cdot 10 = 100$; $10^1 = 10$.

In extending the use of exponents to include zero and the negative integers as well as the positive integers, we define

$$a^0 = 1 \quad (a \neq 0) \qquad \text{and} \qquad a^{-k} = \frac{1}{a^k} \quad (a \neq 0).$$

Note that without the restriction, $a \neq 0$, we would have the possibility of an indicated division by zero, and, as we shall see in Section 4.5 division by zero is not defined. For example, 0^{-3} would result in the following:

$$0^{-3} = \frac{1}{0^3} = \frac{1}{0}.$$

Example 3. $7^0 = 1$; $10^0 = 1$; $6^{-2} = \frac{1}{6^2} = \frac{1}{36}$; $10^{-3} = \frac{1}{10^3} = \frac{1}{1000}$;

and $10^{-1} = \frac{1}{10^1} = \frac{1}{10}$.

We are now prepared to discuss the Hindu-Arabic system of numeration.

2.10   The Hindu-Arabic Numeral System

The system of numeration that is most commonly used is the *Hindu-Arabic system*. It is also referred to as the *decimal system* because of the unique role of ten in its structure (*decem* is the Latin word for ten).

Like most numeral systems, the Hindu-Arabic system went through a long metamorphic period. Although the Hindus are given much of the credit for developing the system, some of its features apparently originated with other peoples. In writings prior to the time of Christ there is no evidence that the Hindus used either zero or place value. It is possible that they obtained the idea of the zero symbol from the Greeks and the concept of place value from the Babylonians. In any event, the Hindus eventually incorporated both features into their numeral system, which was then carried into Europe, probably by traders and other travelers. Historical records indicate that the Arabs invaded North Africa and Spain and brought the Hindu numerals with them. The symbols gradually changed from generation to generation, but with the invention of the printing press in the fifteenth century they became fairly well standardized. Hence, the symbols we work with today are much like those used in the fifteenth and sixteenth centuries. Some variations, however, may be seen on bank checks and on the ticker tape in a stockbroker's office. Perhaps electronic devices will cause another modification of our numerals.

The ten basic symbols in our decimal system of notation are 0, 1, 2, 3, 4, 5, 6, 7, 8, and 9. Each of these symbols is called a digit. Digits are used separately or in combination with other digits to form numerals, just as letters are used separately or in combination with other letters to form words. For example, the numeral 6252 has four digits.

The decimal system has a *place-value property* such that the value of each digit in a numeral is multiplied by some power of ten according to the position of the digit in relation to a *reference point*. This reference point is called the **decimal point** of the numeral. The value of each digit is multiplied by the appropriate power of ten, and these terms are then added together. This means that the system also has an *additive property*.

Example 1. $2059.74 = 2 \cdot 10^3 + 0 \cdot 10^2 + 5 \cdot 10^1 + 9 \cdot 10^0 + 7 \cdot 10^{-1} + 4 \cdot 10^{-2}$. Note that the reference (decimal) point is indicated by a dot on the line between the digits 9 and 7. The power to which ten is raised begins with zero for the digit immediately to the left of the decimal point and increases one unit for each digit as we progress to the left. As we progress to the right, the exponent of ten decreases one unit for each digit. The value of the entire numeral is then found by adding the terms as indicated.

A numeral such as 2059.74 is said to be expressed in **standard form**. In the equation of Example 1, the numeral 2059.74 has been rewritten in what is called **expanded form**. Since $10^3 = 1000$, $10^2 = 100$, and so forth, we could have used any one of several expanded forms of notation such as the following:

$$\begin{aligned}
2059.74 &= 2000 + 0 + 50 + 9 + .7 + .04 \\
&= 2(1000) + 0(100) + 5(10) + 9(1) + 7(.1) + 4(.01) \\
&= 2(1000) + 0(100) + 5(10) + 9(1) + 7(\tfrac{1}{10}) + 4(\tfrac{1}{100}) \\
&= 2(10^3) + 0(10^2) + 5(10^1) + 9(10^0) + 7(10^{-1}) + 4(10^{-2}).
\end{aligned}$$

Observe that if there is no operation symbol between two numerals, the numbers they represent are to be treated as factors and multiplied. For example, $2(1000) = 2 \times 1000 = 2 \cdot 1000 = 2000$.

Actually, the Hindu-Arabic system of numeration does have a multiplicative feature, but the factor by which any given digit is multiplied is (as explained above) determined by the *place* of the digit in relation to the decimal point. Consequently, the best name for this property seems to be *place value*. Recall that the Roman system has a multiplicative feature not determined by the location of a symbol but rather by the placement of one or two bars over a given symbol. Since the Roman system has no symbol for zero, the Romans could not use a place value system in the same way we do. How could the numeral for five hundred six (506) be written using a place value system if we had no symbol for zero?

EXERCISE 2.10

1. (a) What is a digit?
   (b) What are the digits of the Hindu-Arabic numeral system?
2. What is a numeral?

3. (a) Why is the Hindu-Arabic system of numeration sometimes referred to as the decimal system?
  (b) What is meant when we say that the Hindu-Arabic numeral system has a place value property?
  (c) Why is a numeral for zero important in a numeral system having a place value property?

4. What is the purpose of a decimal point in a numeral?

5. Write the standard numeral for each of the following:
  (a) $4(10^3) + 3(10^2) + 7(10^1) + 6(10^0)$
  (b) $5(10^5) + 6(10^3) + 3(10^0) + 2(10^{-2})$
  (c) $7(10^{-3}) + 8(10^{-5}) + 9(10^{-7})$
  (d) $3(1000) + 7(10) + 4(\frac{1}{100}) + 3(\frac{1}{1000})$
  (e) $8(100) + 6(1) + 3(0.01) + 2(0.0001)$

6. Rewrite each of the following standard numerals in expanded form using exponential notation:
  (a) 943   (b) 2342   (c) 105
  (d) 94.2   (e) 0.007   (f) 100.234

REVIEW EXERCISES

1. Which of the following sets are well-defined?
  (a) The ten best artists of the nineteenth century.
  (b) Positive integers less than 17.
  (c) Positive integers greater than 17.
  (d) The countries of the world having hot climates.
  (e) The presidents of the United States who held office prior to 1972.

2. Use set-builder notation to indicate each of the following sets:
  (a) $\{a, b, c, d, \ldots, z\}$ (b) $\{1, 2, 3, 4, \ldots, 100\}$

3. Write in words the meaning of the following:
  (a) $6 \in A$ (b) $C = \{x, y, z\}$ (c) $\{x \mid x \notin B\}$
  (d) $F \subseteq G$ (e) $K \subset L$ (f) $A \not\subset B$

4. Name the proper subsets of $\{a, 3, 4\}$.

5. Explain the difference between equal sets and equivalent sets.

6. What standard set is equivalent to $\{a, b, c, d\}$?

7. Write an Egyptian numeral for each of the following:
  (a) 52   (b) 643   (c) 2,400,234

8. Write the Babylonian numeral for each of the following:
  (a) 32   (b) 133   (c) 214

9. Write the Roman numeral for each of the following:
  (a) 79   (b) 243   (c) 1984

10. Rewrite the following standard numerals in expanded form using exponential notation:
  (a) 235   (b) 27,004   (c) 700.0006

M. C. Escher, *Three Spheres I*
1945. Wood-engraving. 28 x 17 cm.
Escher Foundation, Haags Gemeentemuseum, The Hague

3 OPERATIONS ON SETS

3.1 Introduction

In Chapter 2 we used concepts concerning sets to define and identify the whole numbers. In this chapter, we will study *operations* on sets. Later, in Chapter 4, set operations will be used to define operations on the whole numbers.

At this time, we will examine a few *Venn diagrams*. Venn diagrams are very helpful for developing and understanding various set concepts and operations. As we shall see in Section 3.8, these diagrams are also useful in determining the validity of an argument.

3.2 Venn Diagrams and Universal Sets

The diagrams we shall begin to use in this section are named after John Venn (1834–1923), an English logician. Venn first used this type of diagram in his 1881 publication entitled *Symbolic Logic*. The Swiss mathematician Leonard Euler (1707–1783) also used diagrams to illustrate certain principles of logic, and for this reason Venn-type diagrams are also called *Euler circles*. In **Venn diagrams**, elements of a set are represented by that part of a plane inside of or on a simple closed curve. Sometimes specific elements of a set are indicated by letters or other symbols within the simple closed curve.

Example 1. Represent the set *B* as a proper subset of set *A* by using a Venn diagram.

Solution: Any of the diagrams in Figure 3.2.1 would be satisfactory. The important point is that all of region *B* is contained in region *A*, but since *B* is a *proper* subset of *A*, not all of region *A* is contained in region *B*.

FIGURE 3.2.1

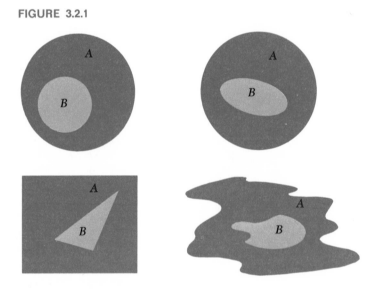

Example 2. Let $G = \{a, b, c\}$ represent members of a basketball team who wear glasses and let $T = \{a, b, c, d, e\}$ represent the team. Illustrate with a Venn diagram.

Solution: The solution is Figure 3.2.2.

Usually it is necessary to restrict the elements from which a given set may be chosen. For example, in a sociological study it might be desired to select a family from a given geographic region, income level, occupation, or racial group. The set of elements from which a given set may be chosen is known as the **universe** or the **universal set** for that particular situation. The latter *U* is generally used to represent the universe unless some other letter is more appropriate.

FIGURE 3.2.2

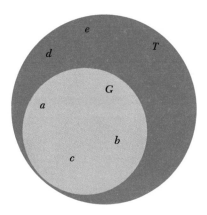

If the basketball team in Example 2 above were selected from Colum-bia College, then the universe would be the students of Columbia College, and the situation could be illustrated by the Venn diagram in Figure 3.2.3.

It can be seen that $G \subset T \subset C$ (the universe). If we wished to consider T as the universe, we could simply omit the rectangle from the diagram and show $G \subset T$. Some authors follow the convention of always represent-ing the universe with a rectangle. If we followed this convention for $G \subset T$, T would be shown as a rectangle.

FIGURE 3.2.3

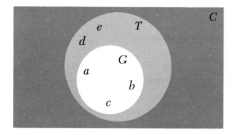

3.3 The Complement of a Set

An important set concept is given in the following definition.

DEFINITION 3.3.1. If U is the universal set and if A is a subset of U, then the set of all elements of U that are not elements of A is the **complement** of A. The complement of A is indicated by A', read "A prime."

Example 1. If $U = \{0, 1, 2, 3, 4\}$ and $A = \{2, 3\}$, then the complement of A is $A' = \{0, 1, 4\}$.

Example 2. Given that the universe T is the set of tenth grade boys in a certain school and that G is the set of those who play the guitar, draw a Venn diagram, shading in G'.

Solution: See Figure 3.3.1.

Example 3. Given that U is the set of all boys in a school, T is the boys in the tenth grade, and G is the tenth grade boys who play the guitar, draw a Venn diagram shading in G'.

Solution: The solution is Figure 3.3.2.

By comparing Examples 2 and 3 we see that the complement of a given set depends on the set chosen as the universe. In Example 2 the complement of G is all the boys in T (the tenth grade) who do not play the guitar, and we say that G' is the complement of G with respect to T. In Example 3 the complement of G is all the boys in U (the school) who do not play the guitar, and we say that G' is the complement of G with respect to U.

FIGURE 3.3.1

FIGURE 3.3.2

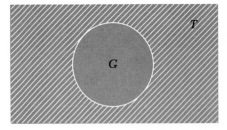

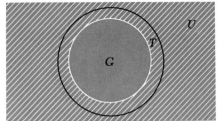

EXERCISE 3.3

1. Draw a Venn diagram showing that:
   (a) The set of all cats is a subset of the set of all animals.
   (b) $\{2, 3, 4\}$ is a subset of $\{2, 3, 4, 5\}$.
   (c) The set of female cats is a subset of all cats, which in turn is a subset of all animals.

2. (a) If the universe is the set of all zebras, what is the complement of the set of female zebras?
   (b) If the universe is the set of all animals, what is the complement of the set of all female zebras?

(c) If the universe is the set of all rabbits, what is the complement of the set of all rabbits?

3. The set of all trout is a subset of the set of all fish, and the set of all fish is a subset of the set of all animals.
   (a) What is the complement of the set of all trout with reference to the set of all fish as the universe?
   (b) What is the complement of the set of all trout if the set of all animals is considered the universe?
   (c) What is the complement of the set of all trout if the set of all trout is the universe?

4. If $U = \{a, b, c, d, e\}$, what is the complement of each of the following sets?

   (a) $\{a, b, c\}$ (b) $\{a\}$
   (c) $\{e, a\}$ (d) $\{b, d, a, c, e\}$
   (e) $\varnothing$ (f) $\{c, a, b\}$

5. Given that U is the universe and $A \subset B \subset U$. Illustrate this with a Venn diagram and shade A' with horizontal parallel lines and B' with vertical parallel lines.

6. (a) What does $(C')'$ equal?
   (b) Simplify $((C')')'$.
   (c) If C is the complement of D, what is the complement of C?

7. If the universe is the set of all computers:
   (a) What is the complement of the set of all computers?
   (b) What is the complement of $\varnothing$?

3.4   The Union of Sets

We are now ready to show how two sets can be used to identify a unique (exactly one) third set. In other words, we shall perform an *operation* on two sets to determine a third set. An operation that is performed on exactly two sets at a time is called a **binary operation**. In Chapter 4, we shall use binary operations on sets to define addition and multiplication, which are binary operations on numbers rather than on sets.

The first binary operation on sets to be discussed is *union*. The symbol for the union of sets is $\cup$ (not to be confused with the capital letter U for the universal set). Union, of course, implies a uniting or joining together.

DEFINITION 3.4.1.   The **union** of two sets A and B is the set containing all elements that belong to A or B. In abbreviated form we write $A \cup B = \{x \,|\, (x \in A) \vee (x \in B)\}$.

In applying the above definition, you should recall that A or B means *either A or B or both*. Note that the operation union is a binary operation (on exactly two sets) and that by using the definition for this operation we identify a unique set associated with A and B. Since $A \cup B$ is a single set, a single letter such as C can be used to name this set and we may write $C = A \cup B$.

Example 1. If $A = \{2, 3, 4\}$ and $B = \{3, 4, 5, 6\}$, then $A \cup B = \{2, 3, 4, 5, 6\}$.

Example 2. If a Venn diagram is used to identify $A \cup B$ in the previous example, we could first shade in A horizontally and then shade in B vertically as in Figure 3.4.1. But, since any shaded region in A *or* B contains elements in $A \cup B$, $A \cup B$ is better shown by making a drawing like Figure 3.4.2, where all the shading is in the same direction. From Figure 3.4.2 we can readily see that $A \cup B = \{2, 3, 4, 5, 6\}$.

FIGURE 3.4.1

FIGURE 3.4.2

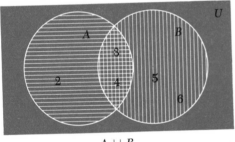

$A \cup B$

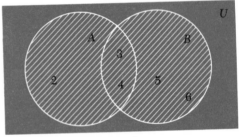

$A \cup B$

3.5 The Intersection of Sets

Another important binary operation on sets is that of *intersection*. The symbol for the intersection of sets is $\cap$.

DEFINITION 3.5.1. The **intersection** of two sets A and B is the set containing all elements that belong to both A and B. In abbreviated form we write $A \cap B = \{x \mid (x \in A) \wedge (x \in B)\}$.

Using the sets that were used in Section 3.4, we shall now identify their intersection rather than their union.

Example 1. If $A = \{2, 3, 4\}$ and $B = \{3, 4, 5, 6\}$, then $A \cap B = \{3, 4\}$.

Example 2. If we were using a Venn diagram to identify $A \cap B$, we could first shade in A with horizontal stripes and then shade in B with vertical stripes as in Figure 3.5.1. The region that is striped in both directions contains elements in $A \cap B$ since it is in A *and* in B. Again we see that $A \cap B = \{3, 4\}$. An alternative diagram is Figure 3.5.2.

FIGURE 3.5.1   **FIGURE 3.5.2**

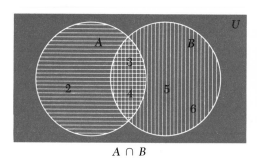

$A \cap B$

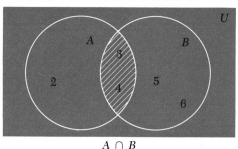

$A \cap B$

Example 3. Given $A = \{a, b, c, d\}$, $B = \{c, d, e\}$, and $U = \{a, b, c, d, e, f, g, h\}$, find $A' \cap B'$ without using Venn diagrams.

Solution: Since $A' = \{e, f, g, h\}$ and $B' = \{a, b, f, g, h\}$, $A' \cap B' = \{e, f, g, h\} \cap \{a, b, f, g, h\} = \{f, g, h\}$.

Example 4. Solve Example 3 using Venn diagrams.

Solution: See Figure 3.5.3. We have represented A' by the region with horizontal parallel lines and B' by the region with vertical parallel lines. Therefore, $A' \cap B'$ is the region with both horizontal and vertical lines and is the set containing three elements $\{f, g, h\}$. A simplified diagram for $A' \cap B'$ is shown in Figure 3.5.4. Observe that $A' \cap B' = (A \cup B)'$.

FIGURE 3.5.3   **FIGURE 3.5.4**

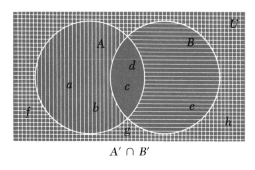

$A' \cap B'$

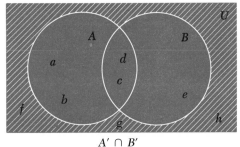

$A' \cap B'$

DEFINITION 3.5.2. If the intersection of two sets A and B is the empty set, then the sets are **disjoint**.

Example 5. If $A = \{a, b, c\}$ and $B = \{d, e\}$, then $A \cap B = \emptyset$ and A and B are said to be disjoint sets. The situation is illustrated by the Venn diagram in Figure 3.5.5, which shows that no region is shared by A and B.

FIGURE 3.5.5

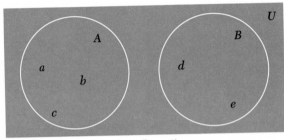

$$A \cap B = \emptyset$$

Although many different binary operations on sets may be defined, we shall confine our discussion to three of the more common and frequently used operations: union and intersection, which have already been defined, and the *Cartesian product*, which is discussed in Section 3.7.

EXERCISE 3.5

1. Define the terms:
   (a) union   (b) intersection
   (c) disjoint

2. What is meant when we say that union and intersection are *binary* operations on sets?

3. Given $A = \{a, b, c, d\}$, $B = \{c, d, e\}$, and $U = \{a, b, c, d, e, f, g\}$, use Venn diagrams to show:
   (a) $A \cup B$ (b) $A \cap B$

4. Using the sets given in Problem 3, identify by the roster method each of the following sets:
   (a) $A \cup B$ (b) $A \cap B$
   (c) A' (d) B'
   (e) $A' \cup B$ (f) $A \cup B'$
   (g) $A' \cap B'$ (h) $A' \cup B'$

5. Given two sets A and B such that $A \cap B \neq \emptyset$, use Venn diagrams similar to Figure 3.5.4 to show:
   (a) A' (b) B'
   (c) $A \cup B$ (d) $A \cap B$
   (e) $A' \cup B$ (f) $A \cup B'$
   (g) $A' \cap B$ (h) $A \cap B'$

6. In Problem 9 of Exercise 1.5, truth tables were used to show that De Morgan's Laws are always true. His Laws may also be illustrated with Venn diagrams. Illustrate that De Morgan's Laws are true by using Venn diagrams to show that:
   (a) $(A \cap B)' = A' \cup B'$
   (b) $(A \cup B)' = A' \cap B'$

7. If the universe G is the set of all girls, H is all girls with blond hair, and E is all girls with blue eyes, describe in words each of the following sets (Venn diagrams may be helpful):
   (a) $H \cup E$ (b) $H \cap E$
   (c) $(H \cap E)'$ (d) $H' \cup E'$
   (e) $(H \cup E)'$ (f) $H' \cap E'$

8. Given that the universe P is the set of all people, R is the set of all redheads, and H is the set of all hot-tempered people, describe in words:
   (a) $R \cup H$ (b) $R \cap H$
   (c) $(R \cup H)'$ (d) $(R \cap H)'$
   (e) $R' \cup H$ (f) $R' \cap H$

3.6   Properties of Union and Intersection

No doubt the reader can think of many binary operations that are defined in mathematics. For example, the addition of numbers is a binary operation in which we associate with two given numbers a unique number called their *sum*. In this chapter, of course, we are dealing with sets rather than numbers and our concern has been with the operations of union and intersection. These operations have some important properties.

(1) $A \cup B = B \cup A$

Perhaps you have already observed that the order in which sets are joined does not affect the result. For example, $\{1, 2, 3\} \cup \{3, 4, 5\} = \{1, 2, 3, 4, 5\}$ and $\{3, 4, 5\} \cup \{1, 2, 3\} = \{1, 2, 3, 4, 5\}$. In general, then, $A \cup B = B \cup A$. Since A and B may be interchanged or *commuted*, we say that the union of sets is commutative.

(2) $A \cap B = B \cap A$

Using the same sets as above, we find that $\{1, 2, 3\} \cap \{3, 4, 5\} = \{3\}$ and that $\{3, 4, 5\} \cap \{1, 2, 3\} = \{3\}$. Although this specific example does not constitute a proof, it is true in general that $A \cap B = B \cap A$. Therefore, we say that the intersection of sets is commutative. Statements (1) and (2) may also be verified by using Venn diagrams (although the use of Venn diagrams is not a general proof either).

(3) $(A \cup B) \cup C = A \cup (B \cup C)$

If three or more sets are to be united, they must be joined two at a time since union is a *binary* operation. We can use various examples or Venn diagrams to convince ourselves that it makes no difference which two sets are united first. In general, then, $(A \cup B) \cup C = A \cup (B \cup C)$, and we say that the union of sets is associative. The parentheses indicate the two sets on which the operation is to be performed first; however, since it makes no difference which two sets are united first, we may write $A \cup B \cup C$ without ambiguity.

Example 1. Let $A = \{2, 3, 4\}$, $B = \{3, 4, 5\}$, and $C = \{5, 6\}$. We shall show that $(A \cup B) \cup C = A \cup (B \cup C)$.

$$\begin{aligned}
(A \cup B) \cup C &= (\{2, 3, 4\} \cup \{3, 4, 5\}) \cup \{5, 6\} \\
&= \{2, 3, 4, 5\} \cup \{5, 6\} \\
&= \{2, 3, 4, 5, 6\} \\
A \cup (B \cup C) &= \{2, 3, 4\} \cup (\{3, 4, 5\} \cup \{5, 6\}) \\
&= \{2, 3, 4\} \cup \{3, 4, 5, 6\} \\
&= \{2, 3, 4, 5, 6\}
\end{aligned}$$

Since $(A \cup B) \cup C$ and $A \cup (B \cup C)$ both equal $\{2, 3, 4, 5, 6\}$, they are names for the same set and we may write $(A \cup B) \cup C = A \cup (B \cup C)$.

(4) $(A \cap B) \cap C = A \cap (B \cap C)$

As with the union of sets, it may be readily verified by an example or by a Venn diagram that the intersection of sets is associative. Symbolically, $(A \cap B) \cap C = A \cap (B \cap C)$. Since it makes no difference on which two sets the operation of intersection is performed first, we may write $A \cap B \cap C$ without ambiguity.

(5) $A \cup (B \cap C) = (A \cup B) \cap (A \cup C)$

(6) $A \cap (B \cup C) = (A \cap B) \cup (A \cap C)$

In referring to (5) and (6), recall once again that the operations within the parentheses are to be performed before the remaining operations. For example, $A \cup (B \cap C)$ means that we must first find the set resulting from the intersection of B and C and then join A with this set.

Note that in (5) and (6) we are dealing with two operations rather than with only one. The properties stated in (5) and (6) are generally known as the *distributive properties*. In (5) we say that union is distributive over intersection and in (6) that intersection is distributive over union. (As an analogy, we have in arithmetic that multiplication is distributive over addition: $2 \cdot (3 + 5) = (2 \cdot 3) + (2 \cdot 5)$; however, addition is not distributive over multiplication: $2 + (3 \cdot 5) \neq (2 + 3) \cdot (2 + 5)$.) Examples 2 and 3 illustrate that intersection is distributive over union.

Example 2. Given $A = \{2, 3, 4\}$, $B = \{3, 4, 5\}$, and $C = \{5, 6\}$, it can be verified that $A \cap (B \cup C) = (A \cap B) \cup (A \cap C)$. (We must say *verify* rather than *prove*, since one example does not constitute a general proof.)

$$A \cap (B \cup C) = \{2, 3, 4\} \cap (\{3, 4, 5\} \cup \{5, 6\})$$
$$= \{2, 3, 4\} \cap \{3, 4, 5, 6\}$$
$$= \{3, 4\}$$

$$(A \cap B) \cup (A \cap C) = (\{2, 3, 4\} \cap \{3, 4, 5\}) \cup (\{2, 3, 4\} \cap \{5, 6\})$$
$$= \{3, 4\} \cup \varnothing$$
$$= \{3, 4\}$$

Since $A \cap (B \cup C)$ and $(A \cap B) \cup (A \cap C)$ are names for the same set $\{3, 4\}$, $A \cap (B \cup C) = (A \cap B) \cup (A \cap C)$.

Example 3. Using Venn diagrams for any sets A, B, and C, it may be shown that $A \cap (B \cup C) = (A \cap B) \cup (A \cap C)$. The left-hand diagram in Figure 3.6.1 illustrates $A \cap (B \cup C)$. The set $(B \cup C)$ is striped horizontally and A is striped vertically. The crosshatched region is contained in A *and* in $(B \cup C)$ and, therefore, represents $A \cap (B \cup C)$. The right-hand diagram of Figure 3.6.1 represents $(A \cap B) \cup (A \cap C)$. We have used horizontal stripes for $(A \cap B)$ and vertical stripes for $(A \cap C)$. The region that is striped horizontally *or* vertically (or both) is in $(A \cap B)$ *or* $(A \cap C)$ and represents $(A \cap B) \cup (A \cap C)$.

FIGURE 3.6.1

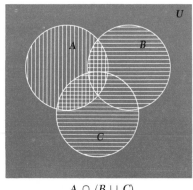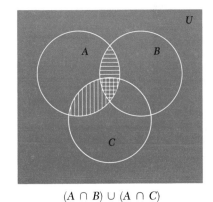

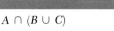

$A \cap (B \cup C)$ $(A \cap B) \cup (A \cap C)$

In summary, then, some of the properties implied by the definitions of the union and intersection of sets are:

1. $A \cup B = B \cup A$ (Union is commutative.)
2. $A \cap B = B \cap A$ (Intersection is commutative.)

3. $(A \cup B) \cup C = A \cup (B \cup C)$ (Union is associative.)
4. $(A \cap B) \cap C = A \cap (B \cap C)$ (Intersection is associative.)
5. $A \cup (B \cap C) = (A \cup B) \cap (A \cup C)$ (Union is distributive over intersection.)
6. $A \cap (B \cup C) = (A \cap B) \cup (A \cap C)$ (Intersection is distributive over union.)

EXERCISE 3.6

1. Given $A = \{a, b, c\}$, $B = \{c, d, e, f\}$, and $C = \{a, e, f, g\}$, use a method similar to Example 2 of this section to verify:
   (a) $A \cup B = B \cup A$
   (b) $A \cap B = B \cap A$
   (c) $A \cap (B \cup C) = (A \cap B) \cup (A \cap C)$
   (d) $A \cup (B \cap C) = (A \cup B) \cap (A \cup C)$

2. Rely on the definitions of union and intersection to complete the following:

ON Test ⟶

   (a) $A \cup \varnothing = ?$ A
   (b) $A \cup A = ?$ A
   (c) $A \cap \varnothing = ?$ $\varnothing$
   (d) $A \cap A = ?$ A
   (e) $A \cap U = ?$ A
   (f) $A \cup U = ?$ U
   (g) $A \cap A' = ?$ $\varnothing$
   (h) $A \cup A' = ?$ U
   (i) If $A \subseteq B$, then $A \cup B = ?$ B
   (j) If $A \subseteq B$, then $A \cap B = ?$ A
   (k) If A and B are disjoint (contain no common elements), then $A \cap B = ?$ $\varnothing$

3. Using Venn diagrams with three intersecting sets similar to those of Figure 3.6.1, verify that $A \cup (B \cap C) = (A \cup B) \cap (A \cup C)$.

4. Use Venn diagrams to determine whether or not $(A \cup B) \cap C$ is equal to $A \cup (B \cap C)$.

5. In a student survey, 16 students were found to enjoy history, 19 liked English, and 18 enjoyed mathematics. History and English were liked by 5, English and mathematics by 8, while 7 enjoyed both history and mathematics. All three disciplines were enjoyed by 3 students and every student liked at least one subject. Beginning with the number of students liking any three subjects, then any two, and so on, use a Venn diagram to determine:

ON Test ⟶

   (a) How many students were in the survey?
   (b) How many students liked English and mathematics but not history?
   (c) How many students liked only English?
   (d) How many students liked history and English but did not enjoy mathematics?
   (e) How many students enjoyed only history?

3.7 Cartesian Products

In order to define the *Cartesian product* of two sets, we must understand what is meant by an ordered pair of elements. From our earlier discussion in Chapter 2, we recall that if a set is identified by listing its elements, order is not important; for example, $\{a, b\} = \{b, a\}$. In some situations, however, the order of elements can be very important. Consider what happens if the numbers 2 and 7 are used in a fraction; clearly, order is important here since $\frac{2}{7} \neq \frac{7}{2}$. Therefore, when we wish to indicate an ordered pair of elements we write (a, b). Note that parentheses (not braces) are used to indicate an ordered pair. In general, $(a, b) = (c, d)$ if and only if $a = c$ and $b = d$.

Once two elements a and b have been ordered, they may be thought of either as two separate elements or as a single element. This would be similar to thinking of a husband and wife as two people or as one married couple. A set whose elements are three ordered pairs may be indicated by writing $\{(a, b), (c, d), (e, f)\}$.

We are now prepared to define the term *Cartesian product*. Cartesian products (sometimes called simply *product sets*) are so named in honor of the famous French mathematician René Descartes (1596–1650).

DEFINITION 3.7.1. The Cartesian product of set A and set B, denoted by $A \times B$, is the set of all ordered pairs (a, b) such that $a \in A$ and $b \in B$.

We may express the above definition in set notation by writing $A \times B = \{(a, b) \,|\, (a \in A) \wedge (b \in B)\}$. Let us point out that the cross symbol $\times$ is used in mathematics in two distinct situations. It is used to indicate the Cartesian product of two sets, and it is also used to indicate the multiplication of two numbers. Hence we may write $A \times B$, read "A cross B," and we may also write 2×7, read "2 times 7." There is little chance for confusion, however, since the context in which the symbol $\times$ is used will clarify its meaning. In Section 4.3 we will define the multiplication of whole numbers in terms of the Cartesian product of sets.

Example 1. If $A = \{a, b, c\}$ and $B = \{2, 3\}$, then $A \times B = \{(a, 2), (a, 3), (b, 2), (b, 3), (c, 2), (c, 3)\}$. The first element of each ordered pair is a member of set A and the second element is a member of set B. Since there are three choices for the first element and two choices for the second, the number of ordered pairs is $3 \times 2 = 6$. Each ordered pair of elements constitutes a *single* element of the Cartesian product set.

Example 2.   Using the sets given in Example 1, we have $B \times A = \{(2, a),$ $(2, b), (2, c), (3, a), (3, b), (3, c)\}$. Again there are six elements in the Cartesian product set but not the same six that were identified as elements of $A \times B$ in Example 1.

The ordered pairs or elements of a Cartesian product set may be found easily by making a table. For any element (a, b), a is called the **first component** and b is called the **second component** of the ordered pair. In a table, it is customary to list the first components horizontally and the second components vertically.

If $A = \{d, e, f\}$ and $B = \{1, 2, 3, 4\}$, the tables for $A \times B$ and $B \times A$ are as illustrated in Figures 3.7.1 and 3.7.2.

FIGURE 3.7.1

FIGURE 3.7.2

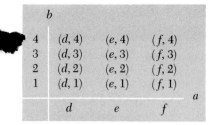

b			
4	$(d, 4)$	$(e, 4)$	$(f, 4)$
3	$(d, 3)$	$(e, 3)$	$(f, 3)$
2	$(d, 2)$	$(e, 2)$	$(f, 2)$
1	$(d, 1)$	$(e, 1)$	$(f, 1)$
	d	e	f	a

$A \times B$

a			
f	$(1, f)$	$(2, f)$	$(3, f)$	$(4, f)$
e	$(1, e)$	$(2, e)$	$(3, e)$	$(4, e)$
d	$(1, d)$	$(2, d)$	$(3, d)$	$(4, d)$
	1	2	3	4	b

$B \times A$

NOT oN Tcs T
EXERCISE 3.7

1. (a) Does $\{a, b\} = \{b, a\}$? Yes
    (b) Does $(a, b) = (b, a)$? No
    (c) Does $\{(a, b), (c, d)\} = \{(c, d), (a, b)\}$? Yes

2. Define *Cartesian product*.

3. Given $A = \{2, 3, 4\}$ and $B = \{7, a\}$:
    (a) Find $A \times B$.
    (b) Find $B \times A$.
    (c) How many elements are in the Cartesian product set $A \times B$?
    (d) How many elements are in the Cartesian product set $B \times A$?
    (e) Is $A \times B = B \times A$? No
    (f) Is $n(A \times B) = n(B \times A)$? Yes

4. Given $C = \{4, 5, 6\}$ and $D = \{5, 6, 7\}$:
    (a) Find $\{(c, d) \mid (c \in C) \wedge (d \in D)\}$.
    (b) Find $\{(d, c) \mid (d \in D) \wedge (c \in C)\}$.
    (c) What is $n(C \times D)$?
    (d) What is $n(D \times C)$?
    (e) How many elements in $C \times D$ are also in $D \times C$?
    (f) Does $n(C) \cdot n(D) = n(C \times D)$?

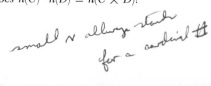

5. Given $A = \{2, 3, 4\}$ and $B = \emptyset$:
  (a) How many elements are in set A?
  (b) How many elements are in set B?
  (c) Find $\{(a, b) | (a \in A) \land (b \in B)\}$.
  (d) Find $\{(b, a) | (b \in B) \land (a \in A)\}$.
  (e) How many elements are in $A \times B$?
  (f) How many elements are in $B \times A$?

6. If $A = \emptyset$ and $B = \emptyset$, find $A \times B$.

3.8 Venn Diagrams and Valid Arguments

To justify valid arguments in Chapter 1 we employed truth tables. However, many people find that Venn diagrams help them to understand various types of arguments. Let us reconsider, for example, a modus ponens form of argument. If x is a member of the set of all women, then x is a member of the set of all humans. We abbreviate this by $W \rightarrow H$ and draw a Venn diagram showing W as a proper subset of H. If, then, it is true that Diana is a woman, we see that she is a member of W. However, since $W \rightarrow H$ we see that Diana is also a human. The argument may be summarized as follows:

Premises:	(1) If you are a woman, then you are a human.	(1) $W \rightarrow H$
	(2) Diana is a woman.	(2) W
Conclusion:	Diana is a human.	$\therefore H$

The argument is illustrated in the modus ponens diagram of Figure 3.8.1.

FIGURE 3.8.1

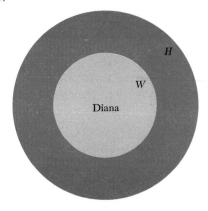

$[(W \rightarrow H) \land W] \rightarrow H$
(Modus ponens)

Consider once again the statement: If x is a member of the set of all women, then x is a member of the set of all humans ($W \rightarrow H$). Suppose Diana is a kitten, then Diana is not human ($\sim H$). Furthermore, we conclude that Diana is not a woman ($\sim W$). This is a modus tollens type of argument and is summarized below:

Premises: (1) If you are a woman,   (1) $W \rightarrow H$
                     then you are a human.
                    (2) Diana is not a human.   (2) $\sim H$

Conclusion: Diana is not a woman. $\therefore \sim W$

The argument is illustrated in the modus tollens diagram of Figure 3.8.2.

FIGURE 3.8.2

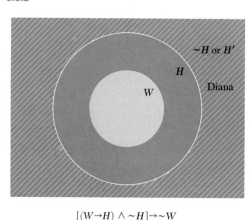

$$[(W \rightarrow H) \wedge \sim H] \rightarrow \sim W$$
(Modus tollens)

Finally, we consider the chain rule (or syllogism). An example follows:

Premises: (1) If you (Diana) are a woman,   (1) $W \rightarrow H$
                     then you are a human.
                    (2) If you are a human,   (2) $H \rightarrow M$
                    then you are a mammal.

Conclusion: If Diana is a woman,
                    then she is a mammal. $\therefore W \rightarrow M$

The argument is illustrated in the chain rule diagram of Figure 3.8.3. It is important to understand that the conclusion in this case is the *entire* implication $W \rightarrow M$, *not* simply the statement M. We have only shown that *if* Diana is a woman, then she is a mammal. Since it may be either true

FIGURE 3.8.3

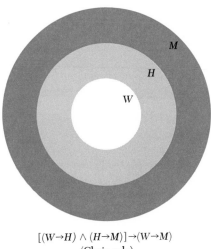

$$[(W{\to}H) \wedge (H{\to}M)]{\to}(W{\to}M)$$
(Chain rule)

or false that Diana is a woman, she is not shown as an element of W in Figure 3.8.3. If it is *true* that Diana is a woman, of course, it is true that she is a mammal since $[(W \to M)$ and $W] \to M$ (modus ponens).

Example 1. Use a Venn diagram to determine whether or not the following is a valid argument: If you work hard, then you will pass this course $(W \to P)$. John does not work hard $(\sim W)$. Therefore, John will not pass this course $(\sim P)$.

Solution: Figure 3.8.4 shows that even though John does not work hard, he may (or may not) be in the set of students passing the course. Hence the given argument is invalid.

FIGURE 3.8.4

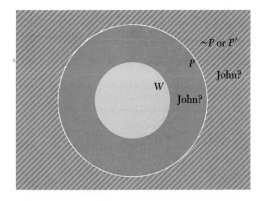

EXERCISE 3.8

1. Use a Venn diagram to test the validity of the argument:
   All clever people like mathematics.
   All people who like mathematics are interesting.
   Therefore:
   T (a) Clever people are interesting.
   F (b) All interesting people are clever.
   T (c) Some interesting people like mathematics.
   F (d) All people who like mathematics are clever.
   T (e) Some interesting people are clever.

2. Test the validity of the following with a Venn diagram:
   All mathematics majors like logic.
   Some girls like logic.
   Some girls are mathematics majors.
   Therefore:
   (a) All mathematics majors are girls.
   (b) All girls who like logic are mathematics majors.
   (c) All girls who are mathematics majors like logic.
   (d) Some girls may neither like logic nor be mathematics majors.
   (e) Some girls are mathematics majors but do not like logic.

3. Use a Venn diagram to test the validity of the argument:
   All girls are beautiful.
   Some girls like English.
   Some girls like history.
   Therefore:
   T (a) Some girls *may* (or may not) like both history and English.
   T (b) Girls who like history are beautiful.
   T (c) Girls who like English are beautiful.
   F (d) All beautiful girls like either history or English.
   F (e) Some girls are not beautiful.

REVIEW EXERCISES

1. If $A = \{1, 2, 3\}$, $B = \{4, 5, 6, 7\}$, and $U = \{1, 2, 3, 4, 5, 6, 7, 8, 9\}$, use the roster method to indicate:
   (a) A' (the complement of A)   (b) B'

2. If the universe is the set of all college students:
   (a) What is the complement of the set of all college students?
   (b) What is the complement of $\emptyset$?

3. If the universe C is the set of all college students, M is all college students enrolled in mathematics, and S is all college students enrolled in sociology, describe in words each of the following sets:

(a) $M \cup S$ (b) $M \cap S$

(c) $(M \cup S)'$ (d) $(M \cap S)'$

4. If the universe $U = \{1, 2, 3, 4, 5, 6, 7, 8\}$, $C = \{1, 2, 3, 4, 5\}$, and $D = \{4, 5, 6, 7, 8\}$, use the roster method to identify each of the following:

(a) $C \cap D$ (b) $C' \cap D$

(c) $(C \cap D)' = C' \cup D'$ (d) $(C \cup D)'$

5. Rely on the definitions of union and intersection to complete the following:

(a) $A \cap A' = ?$ (b) $A \cup A' = ?$

(c) $A \cap \varnothing = ?$ (d) $A \cup \varnothing = ?$

6. Use Venn diagrams with three intersecting sets to verify that $A \cap (B \cup C) = (A \cap B) \cup (A \cap C)$.

7. (a) Does $\{c, d\} = \{d, c\}$?

(b) Does $(c, d) = (d, c)$?

(c) Does $\{(2, 3), (4, 5)\} = \{(4, 5), (2, 3)\}$?

8. Given $A = \{2, 4, 6\}$ and $B = \{c, d\}$, use the roster method to indicate:

(a) $A \times B$ (b) $B \times A$

9. If A and B are sets, under what conditions will $A \times B = B \times A$?

10. Use a Venn diagram to test the validity of the argument:

    All healthy people eat yogurt.

    All people who eat yogurt live a long time.

    Therefore:

    (a) People who are healthy live a long time.

    (b) People who live a long time are healthy.

    (c) Some people who live a long time eat yogurt.

Stuart Davis, *Seme*
1953. Oil. 52 x 40".
The Metropolitan Museum of Art, New York

OPERATIONS ON NUMBERS AND THE USE OF NUMBER BASES

4.1 Introduction

The purpose of this chapter is *not* to teach the reader how to find a sum, such as 642 + 239, or a product, such as 37 × 468; it is assumed that the reader can already determine the sums and products of whole numbers. Instead, our aim is to define various operations on the whole numbers and examine some of the important properties of these operations.

Recall that in Chapter 2 the whole numbers were defined in terms of sets. In this chapter *operations on the whole numbers* will be defined in terms of *operations on sets*. The way in which the operations are defined will determine the properties of these operations. It is the application of these properties along with our system of numeration that enables us to develop efficient methods for finding sums, differences, products, and quotients. Without a knowledge of the properties of number operations, mathematicians could not have developed the familiar techniques and symbolism used in multiplication, long division, and so on. Nor would slide rules and electronic computers exist.

We conclude this chapter by showing how numerals for the whole numbers are written in bases other than ten and how operations on the whole numbers can be performed in various bases. The reader will discover that once he understands the basic process, he will not have to rely on rote memory to perform operations in nondecimal bases.

4.2 Addition in the Whole Numbers

We stated earlier that operations such as union and intersection performed on only two sets at a time are called *binary operations*. Similarly, addition is a binary operation since it is performed on only two numbers at a time.

DEFINITION 4.2.1.   If A and B are disjoint sets $(A \cap B = \emptyset)$ such that $a = n(A)$ and $b = n(B)$, then the **binary operation of addition** $(+)$ assigns to the ordered pair (a, b) a whole number $a + b$ equal to $n(A \cup B)$.

When a and b are added to produce the number $a + b$, a and b are called addends and $a + b$ is called their sum.

Example 1. Suppose A and B are disjoint sets with $A = \{a, b, c\}$ and $B = \{d, e, f, g\}$. If the elements of A and B are counted, we find that $n(A) = 3$ and $n(B) = 4$. But $A \cup B = \{a, b, c, d, e, f, g\}$ and $n(A \cup B) = 7$; therefore, the sum $3 + 4 = 7$ by Definition 4.2.1.

Although the above definition is awkward to apply directly in finding the sum of two large numbers, it serves as the foundation for determining the properties of addition that are discussed next. These properties can then be used to develop more sophisticated means for determining sums, including the design and construction of automatic computers.

For any two elements a and b in the whole numbers W, there *exists* a *unique* (exactly one) element $a + b$ in W, which we call the sum. This is a statement of the closure property for addition in W. We emphasize that closure has two requisites: *existence* and *uniqueness*. By examining the definition of addition in W, we see that since $A \cup B$ exists and is unique, it follows that the sum $a + b$ exists uniquely. For example, $3 + 4$ is the whole number 7 and only 7. The fact that the closure property holds for addition in W is often expressed by saying "the set of whole numbers is closed under addition."

Example 2. Is the set of prime numbers $\{2, 3, 5, 7, 11, 13, 17, \ldots\}$ closed under addition? (A whole number is prime if it has exactly two distinct factors, itself and 1.)

Solution: If the set of prime numbers is closed under addition, the sum of any two primes would have to be another prime. However, $5 + 7 = 12$, and 12 is not prime since it has more than two distinct factors (the factors of 12 are 1, 2, 3, 4, 6, and 12). Therefore, this counterexample allows us to conclude that the closure property does *not* hold for addition in the set of prime numbers. *must say for addition*

The commutative property of addition in W means that for any two elements a and b contained in the set of whole numbers, $a + b = b + a$. By applying Definition 4.2.1 we see that $a + b = n(A \cup B)$ and $b + a = n(B \cup A)$. Since $A \cup B = B \cup A$, it follows that $n(A \cup B) = n(B \cup A)$; hence $a + b = b + a$.

Example 3. Since addition is commutative, $8 + 2 = 2 + 8$; but $8 \div 2 \neq 2 \div 8$ and we see that division is *not* commutative.

Suppose we wish to add more than two numbers. Since addition is a binary operation, only two numbers can be added at a time; therefore, to solve this problem we define $a + b + c = (a + b) + c$. The parentheses indicate that we are to first add a and b. The associative property of addition states that for any elements a, b, and c in W, $(a + b) + c = a + (b + c)$. In other words, if we are given three addends, we may initially add either the first two or the last two. The justification for the associative property depends upon the fact that the operation of union is associative for sets: $(A \cup B) \cup C = A \cup (B \cup C)$.

By using both the commutative and the associative properties of addition it can be proved that for any finite number of addends, any two may be added first, then any other addend may be added to this result, and so on, until the final sum is determined.

Example 4. Verify the associative property for addition by showing that $(3 + 4) + 9 = 3 + (4 + 9)$.

Solution:
$(3 + 4) + 9 = 7 + 9 = 16$, and
$3 + (4 + 9) = 3 + 13 = 16$.
Therefore, $(3 + 4) + 9 = 3 + (4 + 9)$.

Example 5. Show by an example that the associative property does *not* hold for subtraction.

Solution:
$(9 - 4) - 2 = 5 - 2 = 3$.
$9 - (4 - 2) = 9 - 2 = 7$.
But $3 \neq 7$. Therefore, $(9 - 4) - 2 \neq 9 - (4 - 2)$.

There exists a unique number in the set of whole numbers such that when it is used as an addend with any other whole number a, the sum is a. This number is 0 (zero) and is called the **identity element for addition** in W. We express this idea by writing:

$$a + 0 = 0 + a = a \text{ for every } a \text{ in } W.$$

We can apply the definition of addition to show that $a + 0 = a$. Let $a = n(A)$; then $a + 0 = n(A) + n(\emptyset) = n(A \cup \emptyset) = n(A) = a$. Therefore, $a + 0 = a$. Utilizing the fact that addition is commutative, we have $a + 0 = 0 + a = a$.

In summary, then, some of the properties implied by the definition of addition in the whole numbers are as follows.

For any a, b, c in W:

1. There is a uniquely determined sum $a + b$ in W. (The whole numbers are closed under addition.)
2. $a + b = b + a$. (Addition is commutative.)
3. $(a + b) + c = a + (b + c)$. (Addition is associative.)
4. $a + 0 = 0 + a = a$. (Zero is the additive identity element.)

EXERCISE 4.2

1. Define the operation of addition for the whole numbers.
2. Suppose $A = \{a, b, c\}$ and $B = \{b, c, d, e\}$. We find $n(A) = 3$ and $n(B) = 4$. Also, $A \cup B = \{a, b, c, d, e\}$ and $n(A \cup B) = 5$. However, $3 + 4 \neq 5$ even though $n(A \cup B) = 5$. Why not?
3. As we have seen, $n(A \cup B) = n(A) + n(B)$ when $A \cap B = \emptyset$.
    (a) Try to find a general formula for $n(A \cup B)$ that holds when $A \cap B \neq \emptyset$. Use a Venn diagram to aid your thinking.
    (b) Does your formula from part (a) hold when $A \cap B = \emptyset$?
4. (a) State the closure property for addition in the whole numbers.
    (b) Given two whole numbers, such as 14,605 and 2,831,469, what does the closure property for addition tell us about their sum?
5. Try several examples and, by inductive reasoning, decide whether you believe the closure property for addition holds for each of the following sets:
    (a) $\{2, 4, 6, 8, 10, \ldots\}$ (b) $\{10, 20, 30, 40, 50, \ldots\}$
    (c) $\{1, 3, 5, 7, 9, \ldots\}$ NO   (d) $\{3, 6, 9, 12, 15, \ldots\}$
6. Given $A = \{a, b\}$ and $B = \{c, d, e, f\}$, use Definition 4.2.1 to prove that $2 + 4 = 4 + 2$.
7. (a) State the associative property of addition in the whole numbers.
    (b) Use the associative property of addition to find the sum $6 + 9 + 4$ in two different ways.
8. Prove that division in the whole numbers is not commutative.

9. What is meant by saying that there is an identity element for addition in the whole numbers? $3 + 0 = 3$

10. Which property (or properties) of addition is exemplified in each of the following:
    (a) $72 + 34 = 34 + 72$
    (b) $(7 + 5) + 2 = 7 + (5 + 2)$
    (c) $(3 + 4) + 7 = (4 + 3) + 7$
    (d) $34{,}920 + 0 = 34{,}920$
    (e) $(3 + 4) + 5 = 3 + (5 + 4)$
    (f) $a + x = x + a$
    (g) $x + 0 = x$
    (h) $(9 + 2) + 3 = 9 + (3 + 2)$

4.3 Multiplication in the Whole Numbers

Most of us have learned that in finding the product of two whole numbers such as 3×4, we may use 4 as an addend three times. Thus $3 \times 4 = 4 + 4 + 4 = 12$. This approach, however, does cause a problem if the first factor happens to be zero. For example, if we have the product $0 \cdot 3$, what does it mean to use 3 as an addend zero times and what number is indicated by so doing?

If we use the somewhat more sophisticated set-theoretic approach to multiplication, as we did with addition, the above problem is avoided. Hence we will define multiplication in terms of the Cartesian product of sets.

DEFINITION 4.3.1. If A and B are sets with $a = n(A)$ and $b = n(B)$, then the binary operation of multiplication($\cdot$ or $\times$) assigns to the ordered pair (a, b) a whole number $a \cdot b$ equal to $n(A \times B)$.

The number $a \cdot b$ is called the product of the factors a and b. It is also said to be a multiple of both a and b. Some alternative ways of writing the product of a and b are: $a \cdot b$, $a \times b$, $a(b)$, $(a)b$, $(a)(b)$, or simply ab. The reader must be careful not to confuse the cross symbol ($\times$) used between *numbers* to indicate the operation of multiplication with the same or similar symbol used between *sets* to indicate a Cartesian product.

In the definition of addition it was necessary that $A \cap B = \varnothing$. This is not required in the definition of multiplication since the number of elements in $A \times B$ is unaffected by whether or not A and B are disjoint sets.

Example 1. Using the sets $A = \{a, b\}$ and $B = \{a, b, c\}$, find the product $2 \cdot 3$.

Solution:

$n(A) = 2$ and $n(B) = 3$.

$2 \cdot 3 = n(A \times B) = n(\{(a, a), (a, b), (a, c), (b, a), (b, b), (b, c)\}) = 6$.

Therefore, $2 \cdot 3 = 6$.

Next, let us show how the Cartesian product approach to multiplication solves the problem of finding the product $0 \cdot 3$.

Example 2. Using the sets $A = \varnothing$ and $B = \{a, b, c\}$, find the product of 0 and 3.

Solution:

$n(A) = 0$ and $n(B) = 3$.

$0 \cdot 3 = n(A \times B) = n(\varnothing \times B) = n(\varnothing) = 0$.

Therefore, $0 \cdot 3 = 0$.

In general, $0 \cdot a = n(\varnothing) \cdot n(A) = n(\varnothing \times A) = n(\varnothing) = 0$, and in the special case when $a = 0$, we get $0 \cdot 0 = n(\varnothing) \cdot n(\varnothing) = n(\varnothing \times \varnothing) = n(\varnothing) = 0$. Similarly, it may be shown that $a \cdot 0 = 0$. Hence we summarize by stating that: $0 \cdot a = a \cdot 0 = 0$ for all a in W.

Undoubtedly, no one would wish to find the product 634×538 by a direct application of the definition of multiplication. However, the definition serves as a foundation for more sophisticated techniques. By using the definition of multiplication and the properties of multiplication discussed below, we can find the product of any two whole numbers quite simply.

The **closure property for multiplication** in W states that for any two elements a and b in W, there exists a unique element $a \cdot b$ in W, which we call the product of a and b. Often, we express the fact that the closure property holds for multiplication in W by saying "the set of whole numbers is closed under multiplication." It is evident from the definition of multiplication that in this case the closure property depends on the *existence* of a *unique* Cartesian product $A \times B$ for any two sets A and B. In other words, we have assumed the closure property for the Cartesian product set $A \times B$.

Example 3. Is the set of prime numbers $\{2, 3, 5, 7, 11, 13, 17, \ldots\}$ closed under multiplication?

Solution: If the set of prime numbers is closed under multiplication, then the product of any two primes must be another prime. However, $2 \cdot 3 = 6$, and 6 is not prime since it has more than two distinct factors. As a result of this counterexample, we can state that the set of prime numbers is *not* closed under multiplication.

Example 4. Is the set of even numbers closed under multiplication?

Solution: Since an even number is a whole number with at least one factor of 2, the product of any two even numbers may be expressed as $(2a) \cdot (2b) = 4ab$, where a and b are in W. But the product $4ab = 2(2ab)$, which has a factor of 2, and hence is an even number. Therefore, the set of even numbers is closed under multiplication.

For any two elements a and b in W, $a \cdot b = b \cdot a$. This is a statement of the **commutative property of multiplication**. By applying Definition 4.3.1 we see that $a \cdot b = n(A \times B)$ and $b \cdot a = n(B \times A)$. Since $A \times B$ and $B \times A$ are equivalent (*not* equal) sets, it follows that $n(A \times B) = n(B \times A)$; hence $a \cdot b = b \cdot a$.

Since multiplication is a binary operation, we shall define $a \cdot b \cdot c$ to be equal to $(a \cdot b) \cdot c$. You will recall that we defined $a + b + c$ to mean the same as $(a + b) + c$ in considering the associative property of addition. Thus, the **associative property of multiplication** states that for any elements a, b, and c in W, $(a \cdot b) \cdot c = a \cdot (b \cdot c)$. It is not difficult to see that: $(a \cdot b) \cdot c = n(A \times B) \cdot n(C) = n[(A \times B) \times C]$, and $a \cdot (b \cdot c) = n(A) \cdot n(B \times C) = n[A \times (B \times C)]$. The associative property of multiplication is, therefore, a consequence of the fact that $n[(A \times B) \times C] = n[A \times (B \times C)]$.

Using the associative and commutative properties of multiplication, we can prove that the product of any finite number of factors may be found by multiplying any two factors initially, and then multiplying this product by any other factor, and so on.

Example 5. Verify the associative property of multiplication by showing that $(3 \cdot 4) \cdot 6 = 3 \cdot (4 \cdot 6)$.

Solution:
$(3 \cdot 4) \cdot 6 = 12 \cdot 6 = 72$, and
$3 \cdot (4 \cdot 6) = 3 \cdot 24 = 72$.
Therefore, $(3 \cdot 4) \cdot 6 = 3 \cdot (4 \cdot 6)$.

There exists a unique number in the set of whole numbers such that when it is used as a factor with any other whole number a, the product is a. This number is 1 (one) and is called the **identity element for multiplication** in W. We express this idea by writing

$$a \cdot 1 = 1 \cdot a = a \text{ for every } a \text{ in W.}$$

By applying the definition of multiplication, it can be shown that $a \cdot 1 = a$. If we assume $a = n(A)$ and let $\{k\}$ be a representative set containing one element, then:

$$a \cdot 1 = n(A \times \{k\}) = n(A) = a.$$

Therefore, $a \cdot 1 = a$, and by using the commutative property of multiplication, we have $a \cdot 1 = 1 \cdot a = a$.

In summary, then, some of the properties implied by the definition of multiplication in the whole numbers are as follows.

For every a, b, and c in W:

1. There is a uniquely determined product $a \cdot b$ in W. (The whole numbers are closed under multiplication.)
2. $a \cdot b = b \cdot a$ (Multiplication is commutative.)
3. $(a \cdot b) \cdot c = a \cdot (b \cdot c)$ (Multiplication is associative.)
4. $a \cdot 1 = 1 \cdot a = a$ (One is the multiplicative identity element.)

4.4 The Distributive Property of Multiplication over Addition in W

The distributive property involves two operations—multiplication and addition. Before we discuss this property, however, it is appropriate to mention that by convention the operations of *multiplication* and *division* take precedence over those of *addition* and *subtraction* unless otherwise indicated. For example, in the expression $2 + 5 \cdot 3$ we multiply before adding and get $2 + 5 \cdot 3 = 2 + 15 = 17$. However, if we specifically desire addition to take precedence over multiplication, we may so indicate with the use of parentheses and write $(2 + 5) \cdot 3 = 7 \cdot 3 = 21$. Similarly, $2 \cdot 5 + 2 \cdot 3 = 10 + 6 = 16$, while $2 \cdot (5 + 2) \cdot 3 = 2 \cdot 7 \cdot 3 = 42$. Note how the precedence of operations applies in the following discussion of the distributive property.

When we say that multiplication is distributive over addition, we mean that for all a, b, and c in W, $a(b + c) = ab + ac$ and $(b + c)a = ba + ca$. The relation $a(b + c) = ab + ac$ is called the left distributive property since the factor a is to the left of the factor $(b + c)$, which contains two addends. Similarly, the right distributive property states that $(b + c)a = ba + ca$. Since both the left and right distributive properties hold for multiplication over addition, we shall, as a matter of convenience, use the expression the distributive property to include both.

To justify the left distributive property for the whole numbers, we can use the fact that the Cartesian product operation is distributive over union. However, since we did not discuss this result in the previous chapter, we will omit a formal proof of the left distributive property. The right distributive property may be easily proved by using the left distributive property and the commutative property for multiplication.

Example 1. Verify the truth of the distributive property of multiplication over addition by showing that $3(4 + 5) = 3 \cdot 4 + 3 \cdot 5$.

Solution:

$3(4 + 5) = 3 \cdot 9 = 27$ and $3 \cdot 4 + 3 \cdot 5 = 12 + 15 = 27$.
Therefore, $3(4 + 5) = 3 \cdot 4 + 3 \cdot 5$.

If $a(b + c) = ab + ac$, then $ab + ac = a(b + c)$. The distributive property is frequently used in this "reverse" way, particularly in simplifying algebraic expressions.

Example 2. Rename $3 \cdot 4 + 3 \cdot 8$ by using the distributive property.

Solution: $3 \cdot 4 + 3 \cdot 8 = 3(4 + 8)$.

Example 3. Rename $ax + bx$ by using the distributive property.

Solution: $ax + bx = (a + b)x$.

Example 4. Rename $2ax + bx + kx$ by using the distributive property.

Solution: $2ax + bx + kx = (2ax + bx) + kx = x(2a + b) + kx = x[(2a + b) + k] = x(2a + b + k)$. All intermediate steps may be omitted and we can immediately obtain $2ax + bx + kx = x(2a + b + k)$.

EXERCISE 4.4

1. (a) Define multiplication, using the Cartesian product approach.
   (b) Using the definition of multiplication and the sets $A = \{a, b, c, d\}$ and $B = \{1, 2\}$, find the product $4 \cdot 2$ by counting the elements in the product set $A \times B$.

2. Using set $A = \{a, b, c\}$ and set $B = \emptyset$, show that $3 \cdot 0 = 0$.

3. (a) State the closure property for multiplication in the whole numbers.
   (b) Given two whole numbers, such as 432 and 6297, what does the closure property for multiplication tell us about their product? *it's a whole number*

4. Try several examples and then use inductive reasoning to decide whether you believe the closure property of multiplication holds for each of the following sets:
   (a) $\{2, 4, 6, 8, 10, \ldots\}$ *yes*   (b) $\{10, 20, 30, 40, 50, \ldots\}$ *yes*
   (c) $\{1, 3, 5, 7, 9, \ldots\}$ *yes*   (d) $\{1, 2, 4, 8, 16, 32, \ldots\}$ *yes*
   *(e) $\{0, 1\}$ *yes*   *(f) $\{0, 1, 2\}$ *no*
                                      2×2

7(A+3) — mult over addition,

5. If $ab = 0$, what conclusions can you make concerning possible values of a and b?   1 or both must be 0,

6. (a) Using $A = \{a, b\}$ and $B = \{1, 2, 3\}$, verify that $2 \cdot 3 = 3 \cdot 2$ by showing $n(A \times B) = n(B \times A)$.

   (b) Is $A \times B = B \times A$? No

7. (a) State the associative property for multiplication in W.

   (b) Use the associative property of multiplication to find the product $7 \cdot 9 \cdot 3$ in two different ways.

8. In each of the following, one or more of the properties of addition or multiplication is exemplified. Tell which property (or properties) is being applied:

   (a) $3 + 4 = 4 + 3$ Comm of add

   (b) $3 \times 27 = 27 \times 3$ of mult

   (c) $xy = yx$ of mult

   (d) $(3 + 5) + 5 = 3 + (5 + 5)$ ass of add

   (e) $6(4 + 5) = (4 + 5)6$ com of mult

   (f) $8 + n = n + 8$ comm of add.

   (g) $a + (b + 8) = a + (8 + b)$ com + ass of add.

   (h) $(xy)z = x(yz)$ ass of mult

   (i) $(a + x) + 2 = a + (x + 2)$ comm of add.

   (j) $7[(4 + 5) + 6] = 7[4 + (5 + 6)]$ ass of ad.

9. What is meant by saying there is an identity element for multiplication in W?

10. (a) Verify the left distributive property by showing that $6(8 + 3) = 6 \cdot 8 + 6 \cdot 3$.

   (b) Using the same factors, verify the right distributive property.

11. Explain how the distributive property is used in finding the product 2×34.

12. Rename each of the following by using the distributive property. (There may be more than one possible result.)

   (a) $7(a + n)$ (b) $(x + y)a$

   (c) $ax + ay$ (d) $a \cdot 10^2 + 10$ 100A + 10 $100\left(A + \frac{1}{10}\right)$

   (e) $k^2x + k^2$ (f) $4xay + kx$

   (g) $ax + ay + az$ (h) $6 + 9 + 12$

   (i) $ax + a$ (j) $ax^3 + x^3 + bx^3 + cx^3$ $x^3(A + 1 + b + c)$

13. (a) Use the distributive property to show that $7x + 3x = 10x$. $(7+3)x = 10x$

   (b) Use the distributive property to show that $3x + x = 4x$. $10x = 10x$

4.5   Subtraction and Division in the Whole Numbers

Since the operation of subtraction may be defined in terms of addition, it is, from this point of view, not considered a *fundamental* operation. We have the following definition of **subtraction**.

DEFINITION 4.5.1.   If a, b, and k are in W, then $a - b = k$ if and only if $a = b + k$.

In this definition of subtraction, the number a is called the **minuend**, b is called the **subtrahend**, and k is the **difference**. Note that the definition is an "if and only if" statement. This means, for example, that *if* $7 - 2 = 5$, *then* $7 = 2 + 5$; and it also means that *if* $7 = 2 + 5$, *then* $7 - 2 = 5$.

In attempting to find the difference $3 - 9 = k$, we must find a whole number k such that $3 = 9 + k$. However, $3 = 9 + k$ does not hold for any k in W; hence $3 - 9 = k$ has no solution in the set of whole numbers. Since the operation of subtraction does not produce a whole number for *all* ordered pairs (a, b) in $W \times W$, the closure property does *not* hold for subtraction in W.

An example such as $5 - 3 \neq 3 - 5$ shows us that, in general, $a - b \neq b - a$ and subtraction is *not* commutative. Nor does the associative property hold for subtraction in W, for, in general, $(a - b) - c \neq a - (b - c)$. Clearly, $(8 - 5) - 2 \neq 8 - (5 - 2)$ since $(8 - 5) - 2 = 3 - 2 = 1$, while $8 - (5 - 2) = 8 - 3 = 5$.

With the properties of closure, commutativity, and associativity failing for subtraction in the whole numbers, it is somewhat refreshing to find that **multiplication is distributive over subtraction**. We may express this property by writing $a(b - c) = ab - ac$ and $(b - c)a = ba - ca$.

Example 1.   Verify that the left distributive property holds for multiplication over subtraction by showing that $7(9 - 4) = 7 \cdot 9 - 7 \cdot 4$.

Solution:

$7(9 - 4) = 7 \cdot 5 = 35$ and $7 \cdot 9 - 7 \cdot 4 = 63 - 28 = 35$.
Therefore, $7(9 - 4) = 7 \cdot 9 - 7 \cdot 4$.

The reader should be able to verify that the right distributive property holds for multiplication over subtraction.

As with subtraction, division may be defined in terms of another operation and, for this reason, is not considered a fundamental operation. While subtraction is defined in terms of addition, **division** is defined in terms of multiplication.

DEFINITION 4.5.2.   If a, b, and q are in W, and $b \neq 0$, $a \div b = q$ if and only if $a = bq$.

The expression $a \div b = q$ is read, "a divided by b is equal to q"; a is called the **dividend**, b the **divisor**, and q the **quotient**.

The restriction $b \neq 0$ in Definition 4.5.2 is quite necessary. First consider the case where $b = 0$ and $a \neq 0$. For example, when $a = 6$ we have $6 \div 0 = q$. By the definition of division it is necessary that $6 = 0 \cdot q$ if the quotient q exists. However, there is no value for q satisfying this equation since $0 \cdot q = 0$ for all q, and it follows that $a \div 0$ is undefined for all $a \neq 0$.

Now suppose $a = 0$ and $b = 0$. Then $a \div b = 0 \div 0 = q$, and by the definition of division it is necessary that $0 = 0 \cdot q$ if q exists. But *all* whole numbers may be used as values of q to satisfy this equation, and thus the expression $0 \div 0$ is not *uniquely* determined. In fact, it could represent any whole number and hence has limited value. We say, therefore, that $a \div b$ is undefined if $b = 0$ and do not permit the operation of division by zero under any conditions. (Note, however, that $0 \div b = 0$ if $b \neq 0$. For example, $0 \div 5 = 0$.)

Frequently, $\frac{a}{b}$ or a/b is used to mean "a divided by b." For example, $8 \div 2 = \frac{8}{2} = 8/2 = 4$. We also say that 2 divides 8 since $8 = 2 \cdot 4$.

In studying the definition of division, observe that q exists for the equation $a = bq$ if and only if a is a multiple of b. For example, $16 = 5q$ has no solution for q in W because the multiples of 5 in W, $\{0, 5, 10, 15, 20, \ldots\}$, do not include the number 16. Since the operation of division does not assign a whole number to *all* ordered pairs of whole numbers, the closure property does *not* hold for division in W. Furthermore, $a \div b \neq b \div a$ except in special cases and hence division is *not* commutative. Is division associative? It is *not*, for in general, $(a \div b) \div c \neq a \div (b \div c)$. As an example, we see that $(12 \div 6) \div 2 = 2 \div 2 = 1$, but $12 \div (6 \div 2) = 12 \div 3 = 4$. In summary, the properties of closure, commutativity, and associativity do not hold for division in the whole numbers.

What is true is that the whole numbers have a **right distributive property of division over addition** and also over **subtraction**. However, there are no corresponding left distributive properties, and even the right distributive properties of division over addition or subtraction may fail since we do not have closure for division.

The right distributive property of division over addition may be expressed as $(a + b) \div c = (a \div c) + (b \div c)$ or as

$$\frac{a + b}{c} = \frac{a}{c} + \frac{b}{c}.$$

The right distributive property of division over subtraction is analogous.

Example 2.   Verify that the right distributive property of division over addition applies in the following case:

$$\frac{8 + 12}{4} = \frac{8}{4} + \frac{12}{4}.$$

Solution:

$$\frac{8 + 12}{4} = \frac{20}{4} = 5 \quad \text{and} \quad \frac{8}{4} + \frac{12}{4} = 2 + 3 = 5.$$

Therefore, $\dfrac{8 + 12}{4} = \dfrac{8}{4} + \dfrac{12}{4}.$

EXERCISE 4.5

1. (a) Define subtraction in the whole numbers.
   (b) If $9 = 3 + 6$, what subtraction fact is determined by the definition of subtraction?

   $A = B + K$
   $A - B = K$
   $9 - 3 = 6$

   (c) What other closely related subtraction fact may be determined if we know that $3 + 6 = 6 + 3$?

   $9 - 6 = 3$

2. Closure does not hold for subtraction in W. What might be done to remedy the situation? *use integers.*

3. Give an example to show that the commutative property does not hold for subtraction. $8 - 5 = 5 - 3$ *Counter example*

4. Give an example to show that subtraction is not associative. $(4 - 2) - 1 = 4 - (2 - 1)$

5. Illustrate by examples that both the left and right distributive properties hold for multiplication over subtraction.

6. (a) Define division in the whole numbers.
   $\frac{2}{0}$ (b) Why do we bar division by zero when the dividend is not zero? *undefined*
   $\frac{0}{0}$ (c) Why do we bar division by zero when the dividend is zero? *indeterminate*
   (d) Under what conditions do we permit division by zero in the set of whole numbers? *NO*

7. Which of the following properties hold for the operation of division in the whole numbers:
   (a) closure
   (b) commutativity
   (c) associativity

8. (a) Does $(6 + 12) \div 2 = (6 \div 2) + (12 \div 2)$? *yes*
   (b) Does $12 \div (2 + 4) = (12 \div 2) + (12 \div 4)$? *no*

9. Give an example using whole numbers in which a *right* distributive property of division over addition is applicable.

   $(12 + 6) \div 4$
   $\frac{12}{4} + \frac{6}{4}$

10. Give an example using whole numbers in which a *right* distributive property of division over subtraction is applicable.

4.6 Numeral Systems with Nondecimal Bases

The Hindu-Arabic or decimal numeral system was discussed in Section 2.10. In this section we show how numerals can be written in bases other than ten, and in sections that follow we will see how operations on the whole numbers can be performed in various bases. Since most adults have mastered computing in base ten to the point where they can perform the processes automatically, the reader may find it enlightening and interesting to compute in some other base where he cannot rely upon rote memory.

We say that our numeration system has a base of ten because the value of each of the digits in a numeral is multiplied by a power of ten: 10^0, 10^1, 10^2, and so forth. Perhaps the only obvious reason for using a base of ten in our system of numeration is that we have ten fingers. If man had eight fingers, we would probably use a base of eight. The Mayan Indians, who once inhabited southeastern Mexico and Central America, had a highly developed civilization long before they were discovered by Europeans in the sixteenth century. Their numeration system used a base of twenty. (Perhaps they used their toes as well as their fingers for counting!)

As noted earlier, the Hindu-Arabic system uses a set of ten basic symbols or digits—zero through nine. There is no need to invent an entirely new symbol for ten since we have a place-value system; we can write ten as 10 using a combination of two symbols from the ten basic symbols. The situation would be similar if we used any base other than ten. With a base of eight, for example, we would need eight basic symbols—zero through seven. In general, n symbols are required for a base of n in a place value system.

If we consider a base of four, the set of basic symbols is $\{0, 1, 2, 3\}$, and we group at four, writing 10 (read as "one four and zero"). Figure 4.6.1 compares counting in base four with counting in base ten. In a base four numeral, the digits are multiplied by increasing powers of four as we progress to the left: $(four)^0$, $(four)^1$, $(four)^2$, and so forth. If there are digits to the right of a reference point, they are similarly multiplied by $(four)^{-1}$, $(four)^{-2}$, $(four)^{-3}$, and so on. In general, for any base n, the place values are powers of n: n^0, n^1, n^2, and so on to the left of the reference point, and n^{-1}, n^{-2}, n^{-3}, and so forth to the right of the reference point. Observe that we are careful not to call the reference point in base four a decimal point, for this would imply a base of ten.

When we are working in base four, we must read 32 as "three fours and two" or as "three-two, base four," not as "thirty-two" (which would mean three tens and two). In order to show that we are using base four in the numeral 32, we can write 32_{four}. If a numeral has no subscript, it is assumed to be written in base ten unless a specific statement is made to the contrary.

By regrouping the elements of a representative set S, it is possible to show that a numeral in a given base is equivalent to some other numeral in another base. Let us look at some examples.

Example 1. Regroup the elements of a representative set S to show that 32_{four} is equivalent to 14_{ten}.

FIGURE 4.6.1

Base Ten		Base Four	
1	one	1	one
2	two	2	two
3	three	3	three
4	four	10	one four
5	five	11	one four and one
6	six	12	one four and two
7	seven	13	one four and three
8	eight	20	two fours
9	nine	21	two fours and one
10	ten (one ten and zero)	22	two fours and two
11	eleven (one ten and one)	23	two fours and three
12	twelve (one ten and two)	30	three fours
13	thirteen (one ten and three)	31	three fours and one
14	fourteen (one ten and four)	32	three fours and two
15	fifteen (one ten and five)	33	three fours and three
16	sixteen (one ten and six)	100	one four-fours
17	seventeen (one ten and seven)	101	one four-fours, zero fours, and one
18	eighteen (one ten and eight)	102	one four-fours, zero fours, and two
19	nineteen (one ten and nine)	103	one four-fours, zero fours, and three
20	twenty (two tens)	110	one four-fours, one four, and zero
21	twenty-one (two tens and one)	111	one four-fours, one four, and one

Solution: (For simplicity, commas between elements are omitted in the sets below).

Set S	$n(S)$
$\{[a\,b\,c\,d][e\,f\,g\,h][i\,j\ k\ l]m\ n\}$	32_{four}
$\{[a\,b\,c\,d\ \ e\,f\,g\,h\ \ i\,j]k\,l\ m\ n\}$	14_{ten}

Usually we refer to the cardinal number of set S as "fourteen." However, another name for this same cardinal number is 32_{four}, and we may write $14_{ten} = 32_{four}$.

Example 2. By regrouping the elements of a representative set S, show that $23_{five} = 16_{seven}$.

Solution:

Set S	$n(S)$
$\{[a\ b\ c\ d\ e][f\ g\ h\ i\ j]k\ l\ m\}$	23_{five}
$\{[a\ b\ c\ d\ e\ \ f\ g]h\ i\ j\ k\ l\ m\}$	16_{seven}

Example 3. By regrouping the elements of a representative set S, show that $14_{ten} = 112_{three}$.

Solution:

	Set S	$n(S)$
	$\{[a\,b\,c\;\;d\,e\,f\;\;g\,h\,i\;\;j]k\,l\,m\,n\}$	14_{ten}
	$\{[(a\,b\,c)(d\,e\,f)(g\,h\,i)](j\,k\,l)m\,n\}$	112_{three}

If we wish to translate a numeral from base ten to some other base, or from some other base to base ten, we can avoid the cumbersome (but meaningful) method of regrouping by using our knowledge of the decimal system. In base four, for example, we know that the first digit to the left of the reference point (perhaps we should call it a "fours point" instead of a "decimal point") is in the units position ($4^0 = 1$). The next digit to the left is in the fours position ($4^1 = 4$); the next, the sixteens position ($4^2 = 16$); the next, the sixty-fours position ($4^3 = 64$); and so on. To the right of the reference point (fours point) would be the digit representing the number of fourths ($4^{-1} = \frac{1}{4}$); the next, the number of sixteenths ($4^{-2} = \frac{1}{16}$); and so on. See Figure 4.6.2.

FIGURE 4.6.2

10^3	10^2	10^1	10^0	10^{-1}	10^{-2}	10^{-3}	} Base four notation
4^3	4^2	4^1	4^0	4^{-1}	4^{-2}	4^{-3}	⎫
64	16	4	1	$\frac{1}{4}$	$\frac{1}{16}$	$\frac{1}{64}$	⎬ Decimal notation

Place Values in Base Four

The preceding paragraph illustrates the fact that the language we use in discussing other bases is still the language of base ten. For example, in the fours system we have no name for the $(four)^3$ place except the decimal name, the sixty-fours place. We are in the habit of thinking in the decimal system; however, if we were to regularly use some other system, we would undoubtedly develop a suitable vocabulary for it.

The examples below illustrate how our knowledge of computation in the decimal system can be used to translate a numeral from some other base to base ten, or vice versa.

Example 4. Write the numeral in base ten that is equivalent to 312_{four}.

Solution:

$$312_{four} = 3 \cdot 4^2 + 1 \cdot 4^1 + 2 \cdot 4^0 \quad \left.\right\} \text{Expanded form}$$
$$= 3(16) + 1(4) + 2(1) \quad \left.\right\} \text{using base ten}$$
$$= 48 + 4 + 2 \quad\quad\quad\quad \left.\right\} \text{or decimal numerals}$$
$$= 54_{ten}$$

Example 5. Translate 2014_{five} into a decimal numeral.

Solution:

$$\left. \begin{aligned} 2014_{five} &= 2 \cdot 5^3 + 0 \cdot 5^2 + 1 \cdot 5^1 + 4 \cdot 5^0 \\ &= 2(125) + 0(25) + 1(5) + 4(1) \\ &= 250 + 0 + 5 + 4 \\ &= 259_{ten} \end{aligned} \right\} \begin{aligned} &\text{Decimal} \\ &\text{numerals} \end{aligned}$$

Example 6. Translate 26_{ten} into a base three numeral.

Solution: We know that $3^0 = 1$, $3^1 = 3$, $3^2 = 9$, $3^3 = 27$, $3^4 = 81$, and so on. The largest of these numbers that is less than or equal to the given number 26_{ten} is 9. This means that the first digit of our numeral in base three will be in the 9, or 3^2, position. So we find the greatest multiple of 3^2 or 9 that is less than 26, the greatest multiple of 3^1 that is less than the remainder, and so on as indicated below:

$$\left. \begin{aligned} 26_{ten} &= 2(3^2) + 8 \\ &= 2(3^2) + 2(3^1) + 2 \\ &= 2(3^2) + 2(3^1) + 2(3^0) \\ &= 222_{three} \end{aligned} \right\} \begin{aligned} &\text{Expanded form} \\ &\text{using base ten} \\ &\text{numerals} \end{aligned}$$

The smallest number we can use for a base in a place-value system is *two*. If we try to use a base of *one*, we simply end up with a tally numeral system! Using a base of two, called a **binary system**, we need a set of only two basic symbols, 0 and 1, as opposed to ten symbols in the decimal system. Of course, this makes many numerals quite long. For example, the binary numeral for 99 is 1100011. The first eleven numerals in base two are: 0, 1, 10, 11, 100, 101, 110, 111, 1000, 1001, 1010.

The invention and increased use of electronic computers have aroused considerable interest in the base two or binary system of numeration. Since an electrical circuit in a computer is either opened or closed, two alternatives are automatically present to represent the digits 0 and 1. Moreover, the dichotomy of an open or closed circuit can be used not only to represent 0 and 1, but also to answer *yes* or *no*, *true* or *false*, and so on. Thus the logic inherent in a wide variety of problems can be preserved in the machine in the same way as 0 and 1, and everything in a problem is reduced to a sequence of two characters.

If a base larger than ten is used in a place-value numeration system, one needs to invent new symbols. Base twelve, for example, would require unique symbols for ten and eleven. We might represent the numerals of base twelve as follows: 0, 1, 2, 3, 4, 5, 6, 7, 8, 9, T, E, 10, 11, 12, 13, 14, 15, 16, 17, 18, 19, 1T, 1E, 20, 21, and so on. In base twelve numerals, of course, each digit is multiplied by the appropriate power of twelve. The base twelve system is known as the **duodecimal system** since the Latin word for twelve is "duodecim," which means two plus ten.

Example 7. Represent 101101_{two} as a numeral in base ten.

Solution:

$$
\begin{aligned}
101101_{two} &= 1 \cdot 2^5 + 0 \cdot 2^4 + 1 \cdot 2^3 + 1 \cdot 2^2 \\
&\quad + 0 \cdot 2^1 + 1 \cdot 2^0 \\
&= 1(32) + 0(16) + 1(8) + 1(4) \\
&\quad + 0(2) + 1(1) \\
&= 32 + 0 + 8 + 4 + 0 + 1 \\
&= 45_{ten}
\end{aligned}
$$

Decimal numerals

Example 8. Rename 87_{ten} as a numeral in base two.

Solution:

$$
\begin{aligned}
87_{ten} &= 1(2^6) + 23 \\
&= 1(2^6) + 0(2^5) + 1(2^4) + 7 \\
&= 1(2^6) + 0(2^5) + 1(2^4) + 0(2^3) \\
&\quad + 1(2^2) + 3 \\
&= 1(2^6) + 0(2^5) + 1(2^4) + 0(2^3) \\
&\quad + 1(2^2) + 1(2^1) + 1 \\
&= 1010111_{two}
\end{aligned}
$$

Decimal numerals

Example 9. Rename $2E7_{twelve}$ as a numeral in base ten.

Solution:

$$
\begin{aligned}
2E7_{twelve} &= 2 \cdot 12^2 + E \cdot 12^1 + 7 \cdot 12^0 \\
&= 2(144) + 11(12) + 7(1) \\
&= 288 + 132 + 7 \\
&= 427_{ten}
\end{aligned}
$$

Decimal numerals

EXERCISE 4.6

1. By regrouping the elements of a representative set S, show that 15_{ten} is equivalent to 21_{seven}.

2. By regrouping the elements of a representative set S, show that 32_{four} is equivalent to 112_{three}.

3. How many basic symbols are needed in a place-value numeral system
   (a) in base seven?
   (b) in base fifteen?
   (c) in base k?

4. Write each of the following numerals in words:
   (a) 43_{seven}
   (b) 265_{seven}
   (c) 304_{seven}

5. Write the numerals from one to fifteen in base seven.

6. Explain why 273 cannot be a numeral in base six.

7. Why does $3_{four} = 3_{five}$, while $12_{four} \neq 12_{five}$?

8. Rename each of the following by using decimal numerals:
   (a) 10_{two} (b) 111_{two}
   (c) 11101_{two} (d) 110101_{two}

9. Rename the following by using a base two numeral system:
   (a) 3   (b) 18
   (c) 99   (d) 125

10. Using a base twelve (duodecimal) system, rename each of the following:
    (a) 17   (b) 38
    (c) 131   (d) 167

11. Rename each of the following using decimal numerals:
    (a) 34_{five} (b) 123_{four}
    (c) 201_{seven} (d) $T0E_{twelve}$

12. Rename the following using base seven numerals:
    (a) 23_{four} (b) 34_{five} (c) 78_{nine}

4.7 Addition and Subtraction in Nondecimal Bases

The algorithms developed for computing in our decimal system of numeration can also be applied to numeration systems with bases other than ten. The word **algorithm** refers to the particular method and symbolism used in performing an operation in mathematics. Many different algorithms have been developed for the same operation, and very often an algorithm is chosen as a matter of convenience or personal preference. The nature and types of algorithms devised for performing operations in any mathematical system are determined primarily by the definitions of the operations, their properties, and the system of numeration that is used. As a result, the algorithms used in our decimal system can also be used in other place-value systems analogous to it.

In Section 4.6, particular attention was arbitrarily given to writing numerals in base four, so we shall begin here by adding in a base four system. If we wished to compute rapidly in a base four system, it would be necessary to memorize the sums of single-digit numbers just as we do when computing in base ten. However, speed is not our objective here;

rather, we wish to obtain an understanding of the basic principles of computation in place-value systems analogous to our own.

A knowledge of the base ten system can be used to find sums of single-digit numbers in base four. For example, suppose we want to find the sum $3_{four} + 2_{four}$. Using base ten we find $3 + 2 = 5$, but we know that "five" is "one four and one"; hence $3_{four} + 2_{four} = 11_{four}$. The sums of other single-digit numbers in base four may be found in a similar way and a base four addition table constructed as shown in Figure 4.7.1.

If we were to use a base of four rather than a base of ten in our numeration system, it would be necessary to memorize only 4^2 or 16 basic

FIGURE 4.7.1

Base Four

+	0	1	2	3
0	0	1	2	3
1	1	2	3	10
2	2	3	10	11
3	3	10	11	12

Base Two

+	0	1
0	0	1
1	1	10

Base Twelve

+	0	1	2	3	4	5	6	7	8	9	T	E
0	0	1	2	3	4	5	6	7	8	9	T	E
1	1	2	3	4	5	6	7	8	9	T	E	10
2	2	3	4	5	6	7	8	9	T	E	10	11
3	3	4	5	6	7	8	9	T	E	10	11	12
4	4	5	6	7	8	9	T	E	10	11	12	13
5	5	6	7	8	9	T	E	10	11	12	13	14
6	6	7	8	9	T	E	10	11	12	13	14	15
7	7	8	9	T	E	10	11	12	13	14	15	16
8	8	9	T	E	10	11	12	13	14	15	16	17
9	9	T	E	10	11	12	13	14	15	16	17	18
T	T	E	10	11	12	13	14	15	16	17	18	19
E	E	10	11	12	13	14	15	16	17	18	19	1T

Addition Tables

addition facts instead of 10^2 or 100. By using the commutative property, of course, the number of memorized facts can be further reduced. How would we fare with base two or base twelve? See Figure 4.7.1.

Once the sums of single-digit numbers in a given base are found, the algorithms developed for base ten can be used to determine any other sums in the given base. By renaming the numbers using base ten, we can check the computation for possible errors as shown below. When you are checking, first write both the addends and their sum in base ten numerals. Then add using the base ten numerals and see if the sum is correct.

Example 1.   Find the sum $13_{four} + 33_{four}$.

Solution:

    Base Four   *Check (Base Ten)*

       1   1
      13   7
      33   15
     112   22

Example 2.   Find the sum $3_{four} + 13_{four} + 32_{four}$.

Solution:

    Base Four   *Check (Base Ten)*

      2   1
      3   3
     13   7
     32   14
    120   24

Example 3.   Find the sum $1011_{two} + 101_{two}$.

Solution:

    Base Two   *Check (Base Ten)*
    1011   11
     101   5
  10000   16

In subtracting in various bases it is important to remember that when it is necessary to rename the minuend (borrow), the base determines the details of the renaming. Suppose, for example, we wish to use base four and find the difference $32_{four} - 13_{four}$. Using the conventional algorithm, we have:

$$32$$
$$\underline{13}$$

In this case renaming is necessary since 3 is greater than 2. If one of the three fours of the minuend is added to the two ones, the result is "one four and two" or 12_{four}. Thus, we have renamed 32_{four} as $(20 + 12)_{four}$:

$$\overset{2}{\cancel{3}}12$$
$$\underline{1\ 3}$$

On subtracting "three" from "one four and two," we get "three" as shown in the completed solution below.

	Base Four	Check (Base Ten)
	$\overset{2}{\cancel{3}}12$	14
	$\underline{1\ 3}$	$\underline{7}$
	$1\ 3$	7

Example 4. Find the difference $3030_{four} - 233_{four}$.

Solution:

	Base Four	Check (Base Ten)
	$\overset{2}{\cancel{3}}\ \overset{3}{\cancel{0}}\ \overset{12}{\cancel{3}}10$	$\overset{1}{\cancel{2}}\ \overset{9}{\cancel{0}}14$
	$2\ 3\ 3$	$4\ 7$
	$\overline{2\ 1\ 3\ 1}$	$\overline{1\ 5\ 7}$

Example 5. Find the difference $642_{twelve} - 5E_{twelve}$.

Solution:

	Base Twelve	Check (Base Ten)
	$\overset{5}{\cancel{6}}\ \overset{13}{\cancel{4}}12$	$\overset{8}{\cancel{9}}11\ 4$
	$5\ E$	$7\ 1$
	$\overline{5\ T\ 3}$	$\overline{8\ 4\ 3}$

EXERCISE 4.7

1. The following numerals are written in base seven. Rename them using base ten notation.
   (a) 14   (b) 25
   (c) 125   (d) 1043
2. Construct an addition table for base seven.

3. The following numerals are in base seven. Find the sums and check by using base ten numerals.

(a) 3
4

(b) 12
34

(c) 1233
435
2001

(d) 26
53
214
326

4. The following are base seven numerals. Find the differences and check by using base ten numerals.

(a) 4
2
2

(b) 31
5
23

(c) 212
54
125

(d) 2002
666

5. Find the base of the numeration system if:
(a) $2 + 2 = 11$
(b) $6 + 5 = 14$
(c) $7 + 9 = 12$
(d) $2 + 3 = 10$
(e) $10 + 10 = 100$
(f) $4 + 4 = 12$

6. Find the base of the numeration system if:
(a) $23 + 12 = 101$
(b) $65 + 55 = 142$
(c) $78 + 38 = 127$
(d) $66 + 66 = 121$

7. Find the indicated sums and differences using the following base two numerals:

(a) 1
$+10$

(b) 11
-11

(c) 101
$+110$

(d) 1011
-101

(e) 11101
$+10111$

8. Find the indicated sums and differences using the following base twelve numerals:

(a) 93
$+7$

(b) 82
$-T$

(c) 642
$+ET$

(d) 30T
$-EE$

(e) T0T
$+T0E$

4.8 Multiplication and Division in Nondecimal Bases

Let us begin by constructing a multiplication table in base four. Once again, our knowledge of base ten will be helpful. See Figure 4.8.1.

The reader may wish to construct multiplication tables in base two and base seven as an aid in the study of some of the examples below.

Example 1. Assuming the numerals are written in base four, find the product of the ordered pair $(23, 31)$. Check by using base ten.

FIGURE 4.8.1

Base Four

·	0	1	2	3
0	0	0	0	0
1	0	1	2	3
2	0	2	10	12
3	0	3	12	21

Multiplication Table

Solution:

Base Four	*Check (Base Ten)*
31	13
23	11
213	13
1220	130
2033	143

To check the calculation, write the original factors and their product in base ten numerals. Then multiply using the base ten numerals and see if the product is correct. Note that the two partial products in the base four multiplication do not equal the corresponding partial products in the base ten translation. Why not?

Example 2. Assuming the numerals are written in base two, find the product $101 \cdot 1101$.

Solution:

Base Two	*Check (Base Ten)*
1101	13
101	5
1101	65
1101	
1000001	

Example 3. Multiply 263_{seven} by 43_{seven}.

Solution:

	Base Seven	Check (Base Ten)
	263	143
	43	31
	1152	143
	1445	429
	15632	4433

Division is a more difficult operation to perform than the other operations that we have discussed. But then, this is true whether a decimal or a nondecimal system of numeration is used. The two examples below illustrate division in base four, one with a single-digit divisor and the other with a two-digit divisor.

Example 4.   Divide 1311_{four} by 3_{four}.

Solution:

$$
\begin{array}{r}
\text{Base Four} \\
213 \\
3\overline{)1311} \\
12 \\
\overline{11} \\
3 \\
\overline{21} \\
21
\end{array}
\qquad
\begin{array}{r}
\text{Check (Base Ten)} \\
39 \\
3\overline{)117} \\
9 \\
\overline{27} \\
27
\end{array}
$$

Example 5.   Divide 2033_{four} by 23_{four}.

Solution:

$$
\begin{array}{r}
\text{Base Four} \\
31 \\
23\overline{)2033} \\
201 \\
\overline{23} \\
23
\end{array}
\qquad
\begin{array}{r}
\text{Check (Base Ten)} \\
13 \\
11\overline{)143} \\
11 \\
\overline{33} \\
33
\end{array}
$$

EXERCISE 4.8

1. Construct a multiplication table for a base two numeration system.
2. Construct a multiplication table for a base five numeration system.

3. Example 1 (pages 95–96) shows two numbers multiplied to produce a product, first using base four numerals and then using base ten numerals. One can see that the factor $31_{four} = 13_{ten}$; also, the product $2033_{four} = 143_{ten}$.
   (a) Referring to the given example, does $213_{four} = 13_{ten}$?
   (b) Does $1220_{four} = 130_{ten}$?
   (c) Explain the reasons for the results in parts (a) and (b).

4. The following numerals are in base five. Find the products and check by using base ten numerals.

   (a) 2   (b) 3   (c) 43   (d) 23   (e) 132
       2   2   3   34   23
   ——   ——   ——   ——   ——

5. Find the base of the numeration system if:
   (a) $2 \times 4 = 13$ (b) $3 \times 4 = 20$ (c) $2 \times 2 = 11$
   (d) $2 \times 7 = 13$ (e) $2 \times 7 = 16$ (f) $2 \times 7 = 10$

6. Write the numeral for the product 2×3 assuming a base of:
   (a) four   (b) five   (c) six   (d) seven
   (e) eight   (f) nine   (g) ten   (h) eleven

7. The following numerals are in base two. Find the products and check by using base ten numerals.

   (a) 11   (b) 101   (c) 1011
        1   11   10
   ——   ——   ——

8. (a) What is the effect of annexing a zero (on the right) to a numeral in base ten?
   (b) What is the effect of annexing two zeros to a numeral in base ten?
   (c) What is the effect of annexing a zero to a numeral in base seven?
   (d) What is the effect of annexing two zeros to a numeral in base seven?
   (e) Change 23_{four} and 230_{four} to numerals in base ten and show that annexing a zero to 23_{four} has multiplied it by four.
   (f) What is the effect of annexing a zero to a numeral in base n $(n = 2, 3, 4, \ldots)$?

9. The following numerals are in base five. Find the quotients and remainders and check by using base ten numerals.

   (a) $3\overline{)124}$ (b) $4\overline{)233}$ (c) $12\overline{)343}$

10. (a) Prove by the use of base ten numerals that the following problem might have been written in either base four or base five notation.

$$\begin{array}{r} 11 \\ 12\overline{)132} \\ 12 \\ \hline 12 \\ 12 \\ \hline \end{array}$$

   (b) What other bases might have been used in the above example?
   (c) Why is three not a possible base in the example?

11. What base is being used in the following problem? _base 6_

$$
\begin{array}{r}
13 \\
12\overline{)200} \\
12 \\
\hline
40 \\
40 \\
\hline
\end{array}
$$

REVIEW EXERCISES

1. Prove by a counterexample that the operation of subtraction in the whole numbers is not associative. $(9-5)-2 \neq 9-(5-2)$

2. Prove by a counterexample that the operation of division in the whole numbers is not commutative. $8 \div 5 \neq 5 \div 8$

3. (a) Under what conditions will $A \cup B = B \cup A$? _All - comm of_
   (b) Under what conditions will $A \times B = B \times A$? _only if $A = B$_
   (c) Under what conditions will $n(A \times B) = n(B \times A)$? _All condition_
   (d) Under what conditions will $n(A \cup B) = n(A) + n(B)$? _Disjoint set_

4. Show by a counterexample that the closure property does not hold in the whole numbers for:
   (a) subtraction
   (b) division $3 \div 8 \notin w$

5. Rename each of the following by using decimal numerals:
   (a) 423_{five}
   (b) 110101_{two}

6. Rename the following decimal numerals by using a base two system:
   (a) 5   (b) 19   (c) 86

7. Find the base of the numeration system if:
   (a) $4 + 3 = 11$ (b) $5 + 5 = 14$
   (c) $7 + 6 = 13$ (d) $4 + 4 + 5 = 21$

8. Find the base of the numeration system if:
   (a) $2 \times 5 = 13$ (b) $4 \times 9 = 36$
   (c) $5 \times 4 = 24$ (d) $2 \times 4 \times 3 = 20$

9. The following numerals are in base four. Find the products and check by using base ten numerals.
   (a) $\begin{array}{r} 23 \\ 2 \\ \hline \end{array}$
   (b) $\begin{array}{r} 132 \\ 23 \\ \hline \end{array}$

10. The following numerals are in base four. Find the quotients and check by using base ten numerals.
    (a) $2\overline{)32}$
    (b) $3\overline{)1101}$

M. C. Escher, *Cubic Space Division*
1952. Lithograph. 266 x 266 cm.
Escher Foundation, Haags Gemeentemuseum, The Hague

5 THE EVOLUTION OF NUMBERS AND NUMBER SYSTEMS

5.1 Introduction

In Chapter 4 we discussed operations on the whole numbers. You will recall that subtraction and division were defined in terms of addition and multiplication, and, from this point of view, are not considered "fundamental" operations. However, subtraction and division are very useful and significant concepts. The fact that closure holds for neither subtraction nor division in the whole numbers raises serious doubts concerning the adequacy of the whole numbers.

By extending the whole numbers to include negative numbers, closure for subtraction can be accomplished; and by extending the system even further to include the rational numbers (or fractions), division by every number except zero is possible. Therefore, one objective of this chapter is to show how the whole numbers can be extended to include the integers and rational numbers. A few concepts concerning the real, complex, and transfinite numbers will also be presented.

5.2   The System of Whole Numbers

A *number system* should not be confused with a *system of numeration*, which is used in naming numbers. A **number system** consists of a set of elements called numbers and of operations defined on these numbers. The system will have various properties in accordance with the ways in which the elements of the system are chosen and the operations of the system are defined.

The set of whole numbers was defined in Chapter 2. Then, in Chapter 4, we defined several operations on the whole numbers and discussed their properties. Hence we are now prepared to identify the *system of whole numbers*. The essential features of the system are summarized below. Since subtraction and division were defined in terms of addition and multiplication, they are not included in the definition of the system of whole numbers.

DEFINITION 5.2.1.   The **system of whole numbers** consists of the set $W = \{0, 1, 2, 3, 4, \ldots\}$ and the binary operations of addition $(+)$ and multiplication $(\cdot)$. The system has the following properties for any a, b, and c in W:

Closure Properties
1. There is a uniquely determined sum $a + b$ in W.
2. There is a uniquely determined product $a \cdot b$ in W.

Commutative Properties
3. $a + b = b + a$
4. $a \cdot b = b \cdot a$

Associative Properties
5. $(a + b) + c = a + (b + c)$
6. $(a \cdot b) \cdot c = a \cdot (b \cdot c)$

Distributive Property of Multiplication over Addition
7. $a \cdot (b + c) = a \cdot b + a \cdot c$ (Left distributive property)
   $(b + c) \cdot a = b \cdot a + c \cdot a$ (Right distributive property)

Identity Elements
8. There exists a unique element 0 such that for any a in W, $a + 0 = 0 + a = a$.
9. There exists a unique element 1 such that for any a in W, $a \cdot 1 = 1 \cdot a = a$.

5.3   The Set of Integers

Suppose a college student earns $15 and then spends $15. Considering only these two transactions, what is his net worth? Obviously, it is zero. If we use the number 15 to represent $15 earned and invent the number -15 (negative fifteen) to represent an expenditure of $15, then, in computing the student's net worth, it is necessary to define $15 + (-15) = 0$.

In a similar way, for each whole number a we can invent a new unique number $-a$ such that $a + (-a) = 0$. Hence, by definition, $1 + (-1) = 0$, $2 + (-2) = 0$, $3 + (-3) = 0$, and so on. Since we wish to preserve the commutative property of addition, it must also be true that $-1 + 1 = 0$, $-2 + 2 = 0$, $-3 + 3 = 0$, and so forth. Note that in writing a negative number, such as -10, the short horizontal line is part of the numeral and should not be confused with the same symbol used for the subtraction operation.

Whenever the sum of two numbers, $a + b$, is equal to zero (the identity element for addition), b is said to be the **additive inverse** of a. Since addition is commutative, we note that if $a + b = 0$, then $b + a = 0$; hence a and b are additive inverses of each other. For example, the additive inverse of 27 is -27, and the additive inverse of -27 is 27. Therefore, we can write either $--27 = 27$ or $-(-27) = 27$. Similarly, $---27 = -27$, $----27 = 27$, and so on. We have assumed, of course, that the additive inverse of every integer is unique.

If we unite this newly created set of negative numbers with the set of whole numbers we have what constitutes the **set of integers**. The integers may be indicated as:

$$J = \{\ldots, -3, -2, -1, 0, 1, 2, 3, \ldots\}$$

or

$$J = \{\ldots, -3, -2, -1\} \cup \{0\} \cup \{1, 2, 3, \ldots\}.$$

The elements of $\{\ldots, -3, -2, -1\}$ are called the **negative integers**, and the elements of $\{1, 2, 3, \ldots\}$, which were previously called the natural numbers N, may also be called the **positive integers**. An alternative way of writing numerals for the positive integers is $\{+1, +2, +3, \ldots\}$; we read $+1$ as "positive one," $+2$ as "positive two," and so on. The integer 0 is neither positive nor negative. Furthermore, since $0 + 0 = 0$, 0 is its own additive inverse. In fact, 0 is the only integer that is its own additive inverse.

Since 0 is its own additive inverse, it follows that -0 and 0 are numerals for the same number, and $-0 = 0$.

In associating the set of integers with points on the number line, we find it convenient to place them in a one-to-one correspondence with equally spaced points as shown below. As a result, we can now identify points on the number line to the left of 0 as well as to the right of 0.

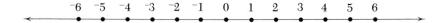

As was true for many other new concepts, negative numbers were not readily accepted. Diophantus of Alexandria (about 250 A.D.) called an equation "absurd" if its roots were negative numbers. Even as late as the seventeenth century, noted mathematicians, such as René Descartes (1596–1650), referred to negative numbers as "false" numbers and avoided their use. It was not until the latter part of the seventeenth century that negative numbers were generally accepted as "true" rather than "false" numbers. Hudde (about 1659) was probably the first algebraist not to offer apologies and misgivings when he allowed letters in equations to represent negative as well as positive numbers. It seems somewhat astonishing that negative numbers, which have so many practical applications, could have existed for hundreds of years without their legitimacy being recognized.

EXERCISE 5.3

1. Briefly describe the meaning of the term *number system*.

2. (a) What two *fundamental* operations are defined on the set of whole numbers?
   (b) Why are subtraction and division not considered fundamental operations in the same sense as addition and multiplication? Refer to Section 5.1 if necessary.

3. Summarize the properties of the system of whole numbers, using a, b, c, and so forth to abbreviate your statements.

4. (a) If you have $10 and spend $15, what integer can be used to represent your net worth?
   (b) If the temperature is 68 degrees and decreases 100 degrees, what integer can be used to represent the temperature?

5. Using the roster method, identify the set of integers J.

6. (a) What is meant by the *additive inverse* of an integer?
   (b) What is the additive inverse of 7?
   (c) What is the additive inverse of -7?

7. What integer is represented by $-x$ if:
   (a) $x = 3$ (b) $x = -4$
   (c) $x = 7$ (d) $x = -7$

⌐ 8. (a) Is 0 a positive number? Explain.
   (b) Is −0 a negative number? Explain.
   (c) How are −0 and 0 related?

5.4 Operations on the Integers

The invention of negative numbers was undoubtedly inspired, to a great extent, by the fact that we did not have closure for subtraction. It would seem that if one has $3 and buys an $8 item, his net worth would be 5 less than zero; or $3 - 8$ should equal "some kind of 5." Should not $3 - 8 = -5$? Then, in the same way that $5 = 5 - 0 = 6 - 1 = 7 - 2 = 8 - 3 = 9 - 4$, and so on, we would have $-5 = 0 - 5 = 1 - 6 = 2 - 7 = 3 - 8 = 4 - 9$, and so on.

By the seemingly innocent act of creating an additive inverse for each of the natural numbers, we accomplish precisely the objectives mentioned in the previous paragraph and obtain closure for subtraction. In fact, by utilizing the additive inverse property we can now define operations on the integers and create the *system of integers*. This system will have all the properties of the whole numbers as well as an additive inverse for each number, which, as previously indicated, results in closure for subtraction.

Before we discuss subtraction, however, let us illustrate how the sum of any two integers may be found by using the additive inverse property.

Example 1.

$$-6 + 8 = -6 + (6 + 2)$$
$$= (-6 + 6) + 2$$
$$= 0 + 2$$
$$= 2$$

Hence, $-6 + 8 = 2$.

Example 2.

$$-4 + (-5) = -4 + (-5) + 0$$
$$= -4 + (-5) + 9 + (-9)$$
$$= -4 + (-5) + (4 + 5) + (-9)$$
$$= (-4 + 4) + (-5 + 5) + -9$$
$$= 0 + 0 + (-9)$$
$$= -9$$

Hence, $-4 + (-5) = -9$.

Using methods similar to those in the above examples, we could prove the following familiar generalizations. Although an example is given to illustrate each generalization, we cannot emphasize too strongly that the example does *not* constitute a *proof* of the generalization.

<table>
<tr><td align="center">*Sum*</td><td align="center">*Example*</td></tr>
<tr><td>$(+a) + (+b) = +(a + b)$</td><td>$(+8) + (+3) = 11$</td></tr>
<tr><td>$(-a) + (-b) = -(a + b)$</td><td>$(-8) + (-3) = -11$</td></tr>
<tr><td>If $a > b$, then</td><td></td></tr>
<tr><td>$(+a) + (-b) = a - b$</td><td>$(+8) + (-3) = 5$</td></tr>
<tr><td>If $b > a$, then</td><td></td></tr>
<tr><td>$(+a) + (-b) = -(b - a)$</td><td>$(+3) + (-8) = -5$</td></tr>
</table>

Subtraction is defined as $a - b = k$ if and only if $a = b + k$. (This is identical to the definition of subtraction in the whole numbers.) However, one can prove that $a = b + k$ if and only if $k = a + (-b)$. We see, then, that the difference of any two integers may be rewritten as a sum, and since we have closure for addition in the integers, we will also have closure for subtraction. This particular property of the integers is cause for the mathematician to rejoice, for, in yielding closure for subtraction, the integers seem to be fulfilling their destiny.

Example 3.   Using the relation $a - b = a + (-b)$ we see, for example, that:

$$(+9) - (+4) = (+9) + (-4) = 5$$
$$(+9) - (-4) = (+9) + (+4) = 13$$
$$(-9) - (+4) = (-9) + (-4) = -13$$
$$(-9) - (-4) = (-9) + (+4) = -5$$

Example 4.   If the temperature is 30 degrees and it falls to -10 degrees, how many degrees did it fall?

Solution: $30 - (-10) = 30 + 10 = 40$. Hence the temperature fell 40 degrees.

Example 5.   If a scuba diver is 40 feet below the water's surface (-40) and then rises to 10 feet below the surface (-10), how far did he rise?

Solution: $-10 - (-40) = -10 + (+40) = 30$. Hence he rose 30 feet.

Although we will not give the proofs here, the uniqueness of the additive inverse of each integer may be used in proving the following relations for multiplication in the integers:

Product	Example
$(+a) \cdot (+b) = +ab$	$(+8) \cdot (+3) = +24$
$(-a) \cdot (-b) = +ab$	$(-8) \cdot (-3) = +24$
$(+a) \cdot (-b) = -ab$	$(+8) \cdot (-3) = -24$
$(-a) \cdot (+b) = -ab$	$(-8) \cdot (+3) = -24$

Students frequently ask why the product of two negative numbers should be a positive number. It is easy to show that this must be the case. For example, we will prove that $(-8) \cdot (-3) = 24$:

1. Using the additive inverse property, we see that $(-8)(-3 + 3) = (-8)(0) = 0$.
2. If the product of step 1 is found by using the distributive property, the result must again be 0. Hence $(-8)(-3 + 3) = (-8)(-3) + (-8)(+3) = (-8)(-3) + (-24) = 0$.
3. However, if $(-8)(-3) + (-24) = 0$ as shown in step 2, it must be true that $(-8)(-3) = 24$. Why? Because the additive inverse of 24 is *unique*.

Finally, let us consider division in the integers. By defining $a \div b = q$ if and only if $a = bq$ where $b \neq 0$, we have the following relations for division in the integers:

Quotient	Example
$\frac{+a}{+b} = +q$	$\frac{+24}{+3} = +8$
$\frac{-a}{-b} = +q$	$\frac{-24}{-3} = +8$
$\frac{+a}{-b} = -q$	$\frac{+24}{-3} = -8$
$\frac{-a}{+b} = -q$	$\frac{-24}{+3} = -8$

Of course, if there is no integer q such that $a = bq$, then the quotient $a \div b$ does not exist in the set of integers. For example, there is no quotient in the integers for $14 \div 3$ since there is no integer q such that $14 = 3q$. Hence we do *not* have closure for the operation of division in the integers.

5.5 The System of Integers

The *system of integers* is essentially an extension of the system of whole numbers. As we have seen, the only basic property of the integers not characteristic of the whole numbers is the existence of an additive inverse for each element in the system. However, as a result of this new property we obtain closure for subtraction; we can identify points on the number line to the left of zero as well as to the right of zero; we have numbers to represent debits as well as credits; we can represent elevations below sea level as well as above sea level, and so on.

The essential features of the system of integers are summarized below. It is interesting to compare Definition 5.5.1 with Definition 5.2.1 for the system of whole numbers.

DEFINITION 5.5.1. The system of integers consists of the set $J = \{\ldots, -3, -2, -1, 0, 1, 2, 3, \ldots\}$ and the binary operations of addition ($+$) and multiplication ($\cdot$). The system has the following properties for any a, b, and c in J:

Closure Properties
1. There is a uniquely determined sum $a + b$ in J.
2. There is a uniquely determined product $a \cdot b$ in J.

Commutative Properties
3. $a + b = b + a$
4. $a \cdot b = b \cdot a$

Associative Properties
5. $(a + b) + c = a + (b + c)$
6. $(a \cdot b) \cdot c = a \cdot (b \cdot c)$

Distributive Property of Multiplication over Addition
7. $a \cdot (b + c) = a \cdot b + a \cdot c$ (Left distributive property)
   $(b + c) \cdot a = b \cdot a + c \cdot a$ (Right distributive property)

Identity Elements
8. There exists a unique element 0 in J such that for any a in J, $a + 0 = 0 + a = a$.
9. There exists a unique element 1 in J such that for any a in J, $a \cdot 1 = 1 \cdot a = a$.

Additive Inverses
10. For each a in J there exists a unique element $-a$ in J such that $a + (-a) = -a + a = 0$.

EXERCISE 5.5

1. What important property does the system of integers possess that the system of whole numbers lacks?

2. What integer is equal to n if:
   (a) $n + 3 = 0$ (b) $n + (-5) = 0$
   (c) $-13 + n = 0$ (d) $8 + n = 0$
   (e) $n + 3 = 2$ (f) $4 + n = 3$

3. Use the additive inverse property of the integers to show that:
   (a) $-6 + 14 = 8$ (b) $-3 + (-4) = -7$

4. Find the following sums:
   (a) $2 + -7 = -5$ (b) $-10 + 3 = -7$
   (c) $-8 + -7 = -15$ (d) $-10 + 0 = -10$
   (e) $-106 + 17 = -89$ (f) $-63 + (-49) = -112$

5. The highest point in California is Mount Whitney, 14,495 feet above sea level, while the lowest point, Death Valley, is 282 feet below sea level. Using subtraction in the integers, find the difference in elevation between these two points.

6. Using the fact that $a - b = a + (-b)$, find the following differences:
   (a) $7 - 9$ (b) $3 - (-8)$
   (c) $5 - (-3)$ (d) $-12 - 4$
   (e) $-8 - 15$ (f) $0 - 3$
   (g) $0 - (-7)$ (h) $7 - 0$

7. Prove that subtraction in J is not commutative by showing that $6 - (-4) \neq -4 - 6$. (This is a proof by counterexample.)

8. Prove that subtraction in J is not associative by showing that $(-8 - 4) - 6 \neq -8 - (4 - 6)$.

9. Express each of the following as an integer:
   (a) $2 \cdot (-3) = -6$ (b) $6(-4) = -24$
   (c) $(-3)(5) = -15$ (d) $(-4)(-5) = 20$
   (e) $-(3 \cdot 4) = -12$ (f) $-(-3 \cdot 4) = 12$
   (g) $-7 \cdot 0 = 0$ (h) $0(-17) = 0$

10. Find $(a)(-b)$ if:
    (a) $a = 2, b = 4$ (b) $a = -5, b = 7$
    (c) $a = 12, b = -1$ (d) $a = 4, b = 0$

11. Prove that $(-3) \cdot (-4) = 12$. [*Hint:* See similar problem in Section 5.4.]

12. Find the following products in two ways:
    (i) Without using the distributive property.
    (ii) Using the distributive property.
    (a) $6(-7 - 4)$ (b) $5(-4 + 9)$

13. Find the following quotients:
    (a) $\frac{8}{-2} = -4$ (b) $\frac{-10}{-5} = 2$ (c) $\frac{0}{-5} =$
    (d) $10 \div (-2) = -5$ (e) $-48 \div (-6) = 8$ (f) $-72 \div 9 = -8$
    (g) $0 \div 10 =$ (h) $\frac{-7}{-1} = 7$ (i) $-1 \div (-1) = 1$

14. Give an example to show that J is not closed with respect to division.

15. How is a related to b if $a \div b = b \div a$?

5.6 The Set of Rational Numbers

By extending the set of whole numbers to include negative integers we obtained closure for subtraction. In a similar manner, the set of integers can be extended so that, with the exception of division by zero, division is always possible. This new set is called the *set of rational numbers* (or *fractions*). In addition to making division possible for all divisors except zero, the rational numbers offer many other advantages. For example, the rational numbers can be associated not only with those points on the number line previously associated with the integers, but also with many other points. As a result, we shall be able to make many more physical measurements of length, area, volume, and so on. (Imagine using only the integers in making a ruler or any other measuring device!)

In considering the set of rational numbers, we shall first identify the elements of the set. A formal definition of a rational number follows.

DEFINITION 5.6.1. A number is **rational** if and only if it can be expressed in the form $\frac{a}{b}$, where a and b are integers and $b \neq 0$.

Observe that the order in which a pair of integers is used to represent a rational number is important since, for example, $\frac{3}{5} \neq \frac{5}{3}$.

When a symbol such as $\frac{3}{4}$ is used to represent a particular rational number, 3 is called the **numerator** and 4 is the **denominator**. The entire expression $\frac{3}{4}$ is frequently referred to as a **fraction numeral** or simply as a **fraction**.

In associating the various fractions with points on the number line, we begin by dividing each unit interval into a given natural number of *congruent* (equal in size and shape) subintervals. For example, if we wish to associate each fraction having a denominator of 4 with a point on the number line, each unit interval on the number line is divided into four congruent subintervals, as shown in Figure 5.6.1. We cannot divide a unit interval into zero subintervals and, hence, never have a denominator of zero.

FIGURE 5.6.1

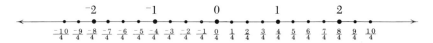

The numerator of the fraction indicates the number of subintervals that must be counted to the right or left of zero in order to identify the point associated with the fraction. For example, the fraction $\frac{3}{4}$ is associated with the point that is three subintervals to the right of zero; $\frac{-3}{4}$ is associated with the point that is three units to the left of zero. If we go neither to the left nor to the right, we are at the point associated with $\frac{0}{4}$. Hence we can have a numerator of zero, even though we never have a denominator of zero.

If, in the example above, each unit interval on the number line is divided into eight congruent subintervals, we discover that $\frac{2}{8}$ is associated with the same point as $\frac{1}{4}$, $\frac{4}{8}$ with the same point as $\frac{2}{4}$, $\frac{6}{8}$ with the same point as $\frac{3}{4}$, and so on. Apparently, then, some fractions are associated with the same point on the number line and can be used to represent the same rational number; we say that such fractions are equal. In general, any two numbers a and b are *equal* if and only if they are names for the same number, and we can write $a = b$. However, the following formal definition of equality for the rational numbers gives us a simple method for determining whether or not two fraction numerals do indeed represent the same rational number.

DEFINITION 5.6.2. If $\frac{a}{b}$ and $\frac{c}{d}$ are rational numbers, then $\frac{a}{b}$ is **equal** to $\frac{c}{d}$ if and only if $ad = bc$.

Example 1. $\frac{2}{17} = \frac{6}{51}$ since $2(51) = 17(6)$.

The *simplest form* (sometimes called the *reduced form*) of a fraction representing a rational number is defined as follows.

DEFINITION 5.6.3. A fraction $\frac{a}{b}$ is said to be in **simplest form** if the greatest common integral divisor of a and b is 1, and b is positive.

If a fraction $\frac{a}{b}$ is in simplest form, then all other fractions representing the same rational number may be generated from it by multiplying both a and b by k, where k is a member of the set of integers and $k \neq 0$. For example, if $\frac{a}{b} = \frac{2}{3}$, then we generate other fractions representing the same rational number as follows:

$$\{\tfrac{2k}{3k} \,|\, (k \in J) \wedge (k \neq 0)\} = \{\ldots, \tfrac{-6}{-9}, \tfrac{-4}{-6}, \tfrac{-2}{-3}, \tfrac{2}{3}, \tfrac{4}{6}, \tfrac{6}{9}, \ldots\}.$$

It is obvious from the preceding example that infinitely many fractions may be used to represent any given rational number.

Figure 5.6.2 illustrates how all fractions representing the same rational number are associated with the same point on the number line. However, while each rational number is associated with exactly one point on the number line, some points on the number line are not associated with any rational number. In fact, infinitely many points on the number line are not associated with any rational number. This fact will be clarified when we discuss the irrational and real numbers.

Before concluding our discussion of the set of rational numbers, we wish to point out that the integers and the rational numbers having denominators of 1 are associated with the same points on the number line. Furthermore, the set of integers and the set of rational numbers with

FIGURE 5.6.2

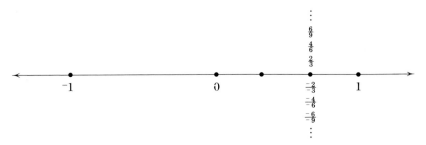

denominators of 1 may be put into a one-to-one correspondence as follows:

$$\ldots \quad \begin{array}{ccccccc} ^-3 & ^-2 & ^-1 & 0 & 1 & 2 & 3 \\ \updownarrow & \updownarrow & \updownarrow & \updownarrow & \updownarrow & \updownarrow & \updownarrow \\ \frac{-3}{1} & \frac{-2}{1} & \frac{-1}{1} & \frac{0}{1} & \frac{1}{1} & \frac{2}{1} & \frac{3}{1} \end{array} \quad \ldots$$

Since each integer a and the corresponding rational number $\frac{a}{1}$ are associated with the same point on the number line and since they also "behave alike" under the operations of addition and multiplication, we will consider that they represent the same number and write $a = \frac{a}{1}$. Thus the integers may be considered to be a subset of the rational numbers. Recall that in a similar way the whole numbers are considered to be a subset of the integers.

EXERCISE 5.6

1. Define a *rational number*.

2. Define the *equals* relation for the rational numbers.

3. Identify three fractions other than $\frac{5}{3}$ that name the rational number five-thirds.

4. How many distinct points on the number line are identified by the fractions $\frac{3}{4}, \frac{2}{3}, \frac{3}{5}, \frac{6}{8}, \frac{6}{9}, \frac{8}{12}, \frac{10}{15}$, and $\frac{1}{2}$?

5. Construct a number line and identify points on it that correspond to the fractions:
   (a) $\frac{2}{5}$ (b) $\frac{-3}{2}$ (c) $\frac{0}{5}$ (d) $\frac{-7}{3}$ (e) $\frac{10}{3}$

6. What fraction is equal to $2\frac{3}{4}$? Explain.

7. What is meant by a fraction in *simplest* or *reduced form*?

8. Express each of the following numbers in simplest form:
   (a) $\frac{4}{8}$ (b) $\frac{-15}{5}$ (c) $\frac{6}{-9}$
   (d) $\frac{0}{4}$ (e) $\frac{2}{-3}$ (f) $\frac{5}{-2}$
   (g) $\frac{51}{17}$ (h) $\frac{-2}{-3}$ (i) $\frac{0}{-1}$

9. Name any five fractions equal to $\frac{0}{1}$.

10. Is $\frac{1}{0}$ a rational number? Explain.

11. Will all fractions equal to $\frac{2}{4}$ be generated by $\{\frac{2k}{4k} | (k \in J) \wedge (k \neq 0)\}$? Explain.

5.7 Operations on the Rational Numbers

If we are to identify the *system* of rational numbers, we must define the operations of addition and multiplication. Since the reader has no doubt studied the operations already, we will immediately define the sum of two rationals as follows.

DEFINITION 5.7.1. If $\frac{a}{b}$ and $\frac{c}{d}$ are any two rational numbers, then their **sum** $\frac{a}{b} + \frac{c}{d} = \frac{ad+bc}{bd}$.

Example 1.

(a) $\frac{4}{5} + \frac{2}{7} = \frac{4 \cdot 7 + 5 \cdot 2}{5 \cdot 7} = \frac{28+10}{35} = \frac{38}{35}$.

(b) $\frac{-4}{3} + \frac{6}{5} = \frac{-4 \cdot 5 + 3 \cdot 6}{3 \cdot 5} = \frac{-20+18}{15} = \frac{-2}{15}$.

By examining Definition 5.7.1 it can be seen that the sum of two rational numbers is defined in terms of multiplication and addition in the integers. Since these operations have closure in the integers, it follows that the sum of any two rational numbers exists uniquely; hence we have closure for addition in the rationals. Using once again the definition of addition in the rational numbers along with the properties of the integers, one can also show that addition in the rationals is both commutative and associative.

The identity element for addition is $\frac{0}{1}$ and for every rational number $\frac{a}{b}$, the sum $\frac{a}{b} + \frac{0}{1} = \frac{0}{1} + \frac{a}{b} = \frac{a}{b}$. The additive inverse of any rational number $\frac{a}{b}$ is $\frac{-a}{b}$ and we see that

$$\frac{a}{b} + \frac{-a}{b} = \frac{ab + (-ab)}{b^2} = \frac{0}{b^2} = \frac{0}{1}.$$

In conclusion, all of the properties that were stated for addition in the integers also hold for addition in the rational numbers.

As in the integers, subtraction can be defined for the rational numbers in terms of addition, and the difference of any two numbers may be rewritten as a sum. Hence we have $\frac{a}{b} - \frac{c}{d} = \frac{a}{b} + \frac{-c}{d}$.

Example 2.

(a) $\frac{6}{7} - \frac{2}{5} = \frac{6}{7} + \frac{-2}{5} = \frac{6(5) + 7(-2)}{7 \cdot 5} = \frac{30 + (-14)}{35} = \frac{16}{35}$.

(b) $\frac{-5}{3} - \frac{-2}{7} = \frac{-5}{3} + \frac{2}{7} = \frac{-5 \cdot 7 + 3 \cdot 2}{21} = \frac{-35 + 6}{21} = \frac{-29}{21}$.

After defining multiplication for the rational numbers, we will observe that not only are all of the properties for multiplication in the integers true for the rationals, but a very important additional property makes possible division by every number except $\frac{0}{1}$. The definition for multiplication follows.

DEFINITION 5.7.2. If $\frac{a}{b}$ and $\frac{c}{d}$ are any two rational numbers, then their **product** $\frac{a}{b} \cdot \frac{c}{d} = \frac{a \cdot c}{b \cdot d}$.

Example 3.
(a) $\frac{3}{5} \cdot \frac{6}{7} = \frac{3 \cdot 6}{5 \cdot 7} = \frac{18}{35}$.
(b) $\frac{6}{7} \cdot \frac{-5}{11} = \frac{6(-5)}{7(11)} = \frac{-30}{77}$.

We see from Definition 5.7.2 that multiplication in the rational numbers is defined in terms of the products of integers. Since closure holds for multiplication in the integers, the rationals will also have closure for multiplication. By continuing to utilize the definition of multiplication in the rationals along with the properties of the integers, it can also be shown that the commutative and associative properties hold for multiplication in the rational numbers. The identity element for multiplication is $\frac{1}{1}$ and for any rational number $\frac{a}{b}$, $\frac{a}{b} \cdot \frac{1}{1} = \frac{1}{1} \cdot \frac{a}{b} = \frac{a}{b}$. We should also observe that both the left and right distributive properties of multiplication over addition hold for the rational numbers.

Finally, we come to the new and important property that is characteristic of the rational numbers but not characteristic of the integers. For each rational number $\frac{a}{b}$, $\frac{a}{b} \neq \frac{0}{1}$, there exists a unique element $\frac{b}{a}$ such that $\frac{a}{b} \cdot \frac{b}{a} = \frac{b}{a} \cdot \frac{a}{b} = \frac{1}{1}$. The element $\frac{b}{a}$ is called the **multiplicative inverse** or the **reciprocal** of $\frac{a}{b}$. Recall that the sum of a number and its additive inverse is equal to the identity element for addition. Similarly, the product of a number and its multiplicative inverse is equal to the identity element for multiplication. It is easy to see that $\frac{a}{b}$ and $\frac{b}{a}$ are each the multiplicative inverse of the other. Why does $\frac{0}{1}$ not have a multiplicative inverse?

After defining division in the rationals, we will show that whether or not division by a given rational number is possible depends on the existence of a multiplicative inverse for that number. Once again division is defined in terms of multiplication.

DEFINITION 5.7.3. If $\frac{a}{b}$, $\frac{c}{d}$, and $\frac{x}{y}$ are rational numbers and $\frac{c}{d} \neq \frac{0}{1}$, then $\frac{a}{b} \div \frac{c}{d} = \frac{x}{y}$ if and only if $\frac{a}{b} = \frac{c}{d} \cdot \frac{x}{y}$.

It would be somewhat tedious to find quotients by the direct application of Definition 5.7.3. However, the definition of division may be used to derive Theorem 5.7.1, which makes it very easy to find quotients in the rational numbers. In the following theorem, and in its proof, we see that division by any given rational number depends on the existence of a multiplicative inverse for that number.

THEOREM 5.7.1. If $\frac{a}{b}$ and $\frac{c}{d}$ are rational numbers with $\frac{c}{d} \neq \frac{0}{1}$, then $\frac{a}{b} \div \frac{c}{d} = \frac{a}{b} \cdot \frac{d}{c}$.

Proof: Let $\frac{a}{b} \div \frac{c}{d} = \frac{x}{y}$. Then, by the definition of division, $\frac{a}{b} = \frac{c}{d} \cdot \frac{x}{y}$. Since products are unique, we may multiply each member of this equation by $\frac{d}{c}$ and obtain $\frac{d}{c}\left(\frac{a}{b}\right) = \frac{d}{c}\left(\frac{c}{d} \cdot \frac{x}{y}\right)$ or $\frac{d}{c} \cdot \frac{a}{b} = \left(\frac{d}{c} \cdot \frac{c}{d}\right)\frac{x}{y}$. Finally, it is seen that $\frac{a}{b} \cdot \frac{d}{c} = \frac{x}{y}$, and we conclude that $\frac{a}{b} \div \frac{c}{d} = \frac{a}{b} \cdot \frac{d}{c}$.

Theorem 5.7.1 justifies the rule that if one rational number is divided by another, the quotient is equal to the dividend multiplied by the reciprocal of the divisor. (Hence the expression, "Invert the divisor and multiply.") Since every rational number except $\frac{0}{1}$ has a reciprocal, Theorem 5.7.1 shows that division in the rationals is always possible except when the divisor is $\frac{0}{1}$.

Example 4.
(a) $\frac{2}{3} \div \frac{3}{4} = \frac{2}{3} \cdot \frac{4}{3} = \frac{8}{9}$.
(b) $\frac{7}{4} \div \frac{-3}{5} = \frac{7}{4} \cdot \frac{5}{-3} = \frac{35}{-12} = \frac{-35}{12}$.
(c) $\frac{0}{1} \div \frac{2}{3} = \frac{0}{1} \cdot \frac{3}{2} = \frac{0}{2} = \frac{0}{1}$.
(d) $\frac{4}{5} \div \frac{0}{1}$ has no solution. Division by zero is undefined.

5.8 The System of Rational Numbers

The essential features of the system of rational numbers have been discussed and are summarized below. You will find it interesting to compare the definition of the system of rational numbers with Definition 5.2.1 for the system of whole numbers and Definition 5.5.1 for the system of integers.

DEFINITION 5.8.1. The **system of rational numbers** consists of the set of all rational numbers Q and the binary operations of addition $(+)$ and multiplication $(\cdot)$ defined on the set Q. The system has the following properties for any $\frac{a}{b}$, $\frac{c}{d}$, and $\frac{e}{f}$ in Q.

Closure Properties
1. There is a uniquely determined sum $\frac{a}{b} + \frac{c}{d} = \frac{ad+bc}{bd}$ in Q.
2. There is a uniquely determined product $\frac{a}{b} \cdot \frac{c}{d} = \frac{ac}{bd}$ in Q.

Commutative Properties
3. $\frac{a}{b} + \frac{c}{d} = \frac{c}{d} + \frac{a}{b}$.
4. $\frac{a}{b} \cdot \frac{c}{d} = \frac{c}{d} \cdot \frac{a}{b}$.

Associative Properties

5. $(\frac{a}{b} + \frac{c}{d}) + \frac{e}{f} = \frac{a}{b} + (\frac{c}{d} + \frac{e}{f})$.

6. $(\frac{a}{b} \cdot \frac{c}{d}) \cdot \frac{e}{f} = \frac{a}{b} \cdot (\frac{c}{d} \cdot \frac{e}{f})$.

Distributive Property of Multiplication over Addition

7. $\frac{a}{b} \cdot (\frac{c}{d} + \frac{e}{f}) = \frac{a}{b} \cdot \frac{c}{d} + \frac{a}{b} \cdot \frac{e}{f}$ (Left distributive property)

$(\frac{c}{d} + \frac{e}{f}) \cdot \frac{a}{b} = \frac{c}{d} \cdot \frac{a}{b} + \frac{e}{f} \cdot \frac{a}{b}$ (Right distributive property)

Identity Elements

8. There exists a unique element $\frac{0}{1}$ such that for any $\frac{a}{b}$ in Q,
$\frac{a}{b} + \frac{0}{1} = \frac{0}{1} + \frac{a}{b} = \frac{a}{b}$.

9. There exists a unique element $\frac{1}{1}$ such that for any $\frac{a}{b}$ in Q,
$\frac{a}{b} \cdot \frac{1}{1} = \frac{1}{1} \cdot \frac{a}{b} = \frac{a}{b}$.

Additive Inverses

10. For each $\frac{a}{b}$ in Q there exists a unique element $\frac{-a}{b}$ in Q such that
$\frac{a}{b} + \frac{-a}{b} = \frac{-a}{b} + \frac{a}{b} = \frac{0}{1}$.

Multiplicative Inverses

11. For each $\frac{a}{b}$ in Q, $\frac{a}{b} \neq \frac{0}{1}$, there exists a unique element $\frac{b}{a}$ in Q such
that $\frac{a}{b} \cdot \frac{b}{a} = \frac{b}{a} \cdot \frac{a}{b} = \frac{1}{1}$.

EXERCISE 5.8

1. Using the definition for the sum of two rational numbers, find the following sums and express each result in simplest form:
   (a) $\frac{2}{3} + \frac{3}{4}$ (b) $\frac{2}{5} + \frac{-1}{2}$ (c) $\frac{7}{3} + \frac{4}{9}$
   (d) $\frac{-3}{2} + \frac{-4}{15}$ (e) $\frac{-2}{3} + \frac{-3}{-8}$ (f) $\frac{7}{6} + \frac{5}{10}$

2. Verify that the addition of rational numbers is commutative by showing that $\frac{2}{7} + \frac{4}{5} = \frac{4}{5} + \frac{2}{7}$.

3. (a) What is the identity element for addition in the integers?
   (b) What is the identity element for addition in the rational numbers?
   (c) What is the sum of $\frac{2}{3}$ and $\frac{0}{-5}$?

4. (a) Why can we say that $\frac{0}{1}$ is in simplest form?
   (b) Are there any other representative elements of zero that are in simplest form? Explain.

5. Express in simplest form the additive inverse of each of the following:
   (a) $\frac{2}{5}$ (b) $\frac{-3}{7}$ (c) $\frac{7}{2}$
   (d) $\frac{a}{b}$ (e) $\frac{2}{-7}$ (f) $\frac{-3}{-2}$

6. Using the equation $\frac{a}{b} - \frac{c}{d} = \frac{a}{b} + \frac{-c}{d}$, find the following differences:
   (a) $\frac{3}{4} - \frac{2}{3}$ (b) $\frac{7}{8} - \frac{-3}{5}$ (c) $\frac{-9}{2} - \frac{7}{8}$
   (d) $\frac{-4}{9} - \frac{-3}{8}$ (e) $\frac{3}{8} - \frac{-7}{3}$ (f) $\frac{13}{4} - 7$

7. (a) Prove by a counterexample that subtraction in the rational numbers is not commutative.
   (b) Prove that subtraction in the rational numbers is not associative by showing that $(\frac{3}{4} - \frac{2}{3}) - \frac{1}{5} \neq \frac{3}{4} - (\frac{2}{3} - \frac{1}{5})$.

8. (a) Verify the commutative property of multiplication using the numbers $\frac{3}{7}$ and $\frac{4}{5}$.
   (b) Verify the associative property of multiplication by showing that
   $(\frac{3}{4} \cdot \frac{5}{7}) \cdot \frac{2}{11} = \frac{3}{4} \cdot (\frac{5}{7} \cdot \frac{2}{11})$.

9. Express in simplest form the multiplicative inverse or reciprocal of each of the following numbers:
   (a) $\frac{3}{4}$ (b) $\frac{-5}{7}$ (c) $\frac{-2}{-3}$
   (d) -5 (e) $\frac{6}{-5}$ (f) $\frac{a}{b}$

10. Using Theorem 5.7.1, find the following quotients:
    (a) $\frac{2}{5} \div \frac{4}{3}$ (b) $\frac{0}{1} \div \frac{-3}{5}$ (c) $\frac{1}{4} \div \frac{1}{4}$
    (d) $\frac{2}{3} \div 5$ (e) $7 \div \frac{3}{4}$ (f) $-3 \div \frac{2}{5}$

11. Prove that division is not associative by showing that
    $(\frac{3}{5} \div \frac{2}{3}) \div \frac{7}{2} \neq \frac{3}{5} \div (\frac{2}{3} \div \frac{7}{2})$.

12. Divide 3 by 8 using the corresponding rational numbers $\frac{3}{1}$ and $\frac{8}{1}$.

13. Using the definition of addition for the rational numbers, prove that
    $\frac{a}{c} + \frac{b}{c} = \frac{a+b}{c}$.

14. (a) If $\frac{0}{c}$ is a rational number, what numbers are permitted as values of c?
    (b) Using the definition of addition for the rational numbers, show that
    $\frac{a}{b} + \frac{0}{c} = \frac{a}{b}$.

5.9   Order and the Denseness Property of the Rationals

Until now we have assumed at least an intuitive notion of what is meant by saying that 6 is less than 7, or -5 is less than -3, and so on. In fact, in associating numbers with points on the number line, we have always associated smaller numbers with points to the left of those points associated with larger numbers.

It seems appropriate now to formally define the *less than* relation. For any of the sets of numbers studied so far, a definition analogous to the following could be used.

DEFINITION 5.9.1. If a and b are rational numbers, a is **less than** b, denoted by $a < b$, if and only if there exists a positive number c such that $a + c = b$.

Example 1.
(a) $6 < 7$, since $6 + 1 = 7$.
(b) $-7 < -5$, since $-7 + 2 = -5$.
(c) $\frac{1}{2} < \frac{3}{4}$, since $\frac{1}{2} + \frac{1}{4} = \frac{3}{4}$.

In attempting to determine the smaller of two numbers we frequently find the application of Definition 5.9.1 cumbersome; for example, $\frac{9}{13} < \frac{7}{10}$ since $\frac{9}{13} + \frac{1}{130} = \frac{7}{10}$. A more practical definition, consistent with Definition

5.9.1, follows. (Note the similarity between Definition 5.9.2 for the less than relation in the rationals and Definition 5.6.2, which defines equal rational numbers.)

DEFINITION 5.9.2. If $\frac{a}{b}$ and $\frac{c}{d}$ are rational numbers with b and d positive integers, then $\frac{a}{b} < \frac{c}{d}$ if and only if $ad < bc$.

Definition 5.9.2 makes it easy to determine whether or not a given rational number is less than some other rational number.

Example 2. $\frac{9}{13} < \frac{7}{10}$ since $9 \cdot 10 < 13 \cdot 7$ or $90 < 91$.

Example 3. Show that $\frac{3}{-7} < \frac{2}{9}$.

Solution: First we rewrite $\frac{3}{-7}$ as $\frac{-3}{7}$. Then $\frac{-3}{7} < \frac{2}{9}$ since $-3(9) < 7(2)$ or $-27 < 14$.

Of course, if $\frac{a}{b} < \frac{c}{d}$, an equivalent statement would be to say that $\frac{c}{d}$ is **greater than** $\frac{a}{b}$, as indicated by $\frac{c}{d} > \frac{a}{b}$.

Example 4. $\frac{3}{4} > \frac{2}{7}$ since $\frac{2}{7} < \frac{3}{4}$.

A number k is said to be **between** a and b if and only if $a < k$ and $k < b$. For example, 8 is between 6 and 9 since $6 < 8$ and $8 < 9$. The expression $a < k$ *and* $k < b$ can be abbreviated by writing $a < k < b$. Hence, for the set of integers we may write:

$$\ldots -3 < -2 < -1 < 0 < 1 < 2 < 3 \ldots$$

As you know from studying the integers, it is not always possible to find an integer between two given integers; for example, there is no integer between 4 and 5. In the system of rational numbers, however, it is *always* possible to find another rational number between any two rational numbers. For example, we can find a number midway between any two distinct rational numbers r_1 and r_2 merely by computing their *arithmetic mean* (or *average*) $M = \dfrac{r_1 + r_2}{2}$. The number midway between $\frac{1}{4}$ and $\frac{1}{2}$ is $(\frac{1}{4} + \frac{1}{2}) \div 2 = \frac{3}{8}$, and we see that $\frac{1}{4} < \frac{3}{8} < \frac{1}{2}$. Similarly, we can find a number between $\frac{1}{4}$ and $\frac{3}{8}$ as well as between $\frac{3}{8}$ and $\frac{1}{2}$. It is evident that by continuing this process infinitely many numbers can be found between the original two numbers $\frac{1}{4}$ and $\frac{1}{2}$. The remarkable fact is that no matter how "close together" any two rational numbers may be, there are infinitely many rational numbers between them. To indicate this special characteristic we say, for obvious reasons, that the rational numbers are **dense** or that they have the property of **denseness.** One would probably guess that since the rationals are so close together, there could be no other numbers between them. However, this is *not* the case, for despite the density of the rationals, there are infinitely many *other* (nonrational) numbers between any two given rationals; these numbers are called *irrational* and are discussed in Section 5.11.

EXERCISE 5.9

1. Use Definition 5.9.1 to prove each of the following:
   (a) $7 < 10$ (b) $-4 < -3$ (c) $-14 < 6$

2. Use Definition 5.9.2 to prove each of the following either true or false. Recall that the denominators of the fractions to be tested must be positive.
   (a) $\frac{2}{3} < \frac{3}{4}$ (b) $\frac{5}{7} < \frac{7}{10}$
   (c) $\frac{9}{8} < \frac{10}{9}$ (d) $\frac{-2}{-3} < \frac{1}{2}$
   (e) $\frac{-2}{3} < \frac{-1}{-2}$ (f) $\frac{-5}{2} < \frac{-2}{3}$
   (g) $\frac{2}{-5} < \frac{-1}{5}$ (h) $-4 < \frac{-1}{2}$

3. (a) For what integral values of x will $\frac{x}{6}$ be equal to $\frac{3}{2}$?
   (b) For what integral values of x will $\frac{x}{6}$ be less than $\frac{3}{2}$?
   (c) For what integral values of x will $\frac{x}{6}$ be greater than $\frac{3}{2}$?

4. Insert the correct relation symbol ($<$, $=$, or $>$) between the numbers in each of the following pairs:
   (a) $\frac{3}{4}, \frac{4}{5}$ (b) $\frac{2}{3}, \frac{7}{11}$
   (c) $\frac{-3}{4}, \frac{2}{3}$ (d) $\frac{25}{32}, \frac{83}{100}$
   (e) $\frac{16}{28}, \frac{11}{18}$ (f) $\frac{-3}{2}, \frac{-14}{9}$

5. What is meant when we say that the rational numbers are *dense*?

6. Find the number corresponding to the midpoint between:
   (a) $\frac{1}{2}$ and $\frac{5}{9}$ (b) $\frac{-3}{4}$ and $\frac{2}{5}$ (c) $\frac{3}{5}$ and 7

7. (a) How many integers are between -4 and 2? Name them.
   (b) How many integers are between 1 and $1,000,000$? Use set notation to identify them.
   (c) How many integers are between 13 and 14?

8. (a) How many rational numbers are between 3 and 4? Use set notation to identify them. Name two of them.
   (b) How many rational numbers are between $\frac{1}{2}$ and $\frac{3}{4}$? Use set notation to identify them. Name two of them.
   (c) How many rational numbers are between $\frac{3}{200}$ and $\frac{4}{200}$? Use set notation to identify them. Name two of them.

5.10 The Pythagorean Theorem

The discussion in Section 5.11 deals with an important set of numbers called the irrational numbers. In identifying some of the irrational numbers, we will find it convenient to use the famous *Pythagorean Theorem;* hence, let us investigate this important theorem now.

The theorem is named after the Greek philosopher and mathematician, Pythagoras, who lived from about 572–501 B.C. Although Pythagoras is usually credited with giving the first general proof, there is evidence that the theorem was known long before he lived. The Chinese referred to the relationships expressed in the theorem in about 1100 B.C., and the Egyptians did so perhaps as early as 2000 B.C.

The Pythagorean Theorem states a relationship that exists among the sides of a right triangle. In a right triangle, the side opposite the right angle is called the *hypotenuse* and the other two sides are called the *legs.* The Pythagorean Theorem says that in any right triangle the square of the length of the hypotenuse is equal to the sum of the squares of the lengths of the legs; that is, if c represents the length of the hypotenuse and if a and b represent the lengths of the legs, then $c^2 = a^2 + b^2$. For example, if the lengths of the legs are 3 and 4, then the length of the hypotenuse must equal 5 since $5^2 = 3^2 + 4^2$. Figure 5.10.1 shows the geometric relationship between the lengths of the sides of a 3, 4, 5 right triangle and the areas of the squares constructed on these sides; the areas are 3^2, 4^2, and 5^2.

FIGURE 5.10.1

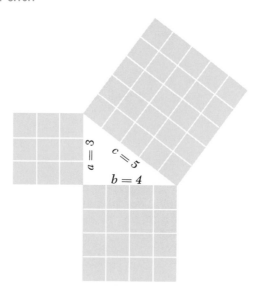

In the above example we see that if the lengths of the legs of a right triangle are known to be 3 and 4, then the length of the hypotenuse may be determined by solving the equation $c^2 = 3^2 + 4^2 = 25$. The number

c is the number that when used as a factor twice will equal 25. Such a number is called the square root of 25, and there are two such numbers: 5 and -5. We are interested in the positive square root of 25 and will indicate it by writing $\sqrt{25} = 5$. The negative square root of 25 is indicated by writing $-\sqrt{25} = -5$. A formal statement of the Pythagorean Theorem follows.

THEOREM 5.10.1.   The Pythagorean Theorem. If a triangle is a right triangle, then the square of the length of the hypotenuse is equal to the sum of the squares of the lengths of the legs.

Although the exact nature of the proof by Pythagoras is not known, many proofs of the Pythagorean Theorem have been presented since his time. It seems likely that Pythagoras' proof was based on the relationship of the areas of three squares rather than simply on a numerical relationship. The proof of the Pythagorean Theorem given below depends upon the concept of area—perhaps it is similar to the original proof of Pythagoras.

In Figure 5.10.2 we see a diagram of a large square, a small square, and four congruent right triangles. Each right triangle has legs with *arbitrary* lengths of a and b. The object is to show that in any of the right triangles, the square of the length of the hypotenuse c is equal to the sum of the squares of the lengths of the legs, or that $c^2 = a^2 + b^2$.

FIGURE 5.10.2

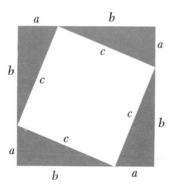

By expressing the area of the large square in two different ways, we can make the following statements, which lead to the desired conclusion:

$$c^2 + 4(\tfrac{1}{2}ab) = (a + b)^2$$
$$c^2 + 2ab = a^2 + 2ab + b^2$$
$$c^2 = a^2 + b^2$$

Literally hundreds of proofs for the Pythagorean Theorem have been devised. Perhaps you would like to try to prove the Pythagorean Theorem by using an original geometric diagram. One proof was invented by a president of the United States, James A. Garfield. It is not surprising that Garfield devised an original proof for the Pythagorean Theorem since he was a well-educated man and enjoyed intellectual activities. He occasionally entertained friends by simultaneously writing a statement in Greek with one hand and in Latin with the other.

5.11   Irrational Numbers

Rational numbers are so named because each is an ordered pair of integers that may be thought of as a *ratio*. Some numbers, however, *cannot* be expressed as the ratio of two integers and are, therefore, called irrational numbers. We will now show how the Pythagorean Theorem may be applied to identify some of the irrational numbers.

Suppose each leg of a given right triangle is of unit length; then $c^2 = 1^2 + 1^2 = 2$ and the length of the hypotenuse is $c = \sqrt{2}$. As we shall soon prove, the number $\sqrt{2}$ cannot be represented as an ordered pair of integers $\frac{a}{b}$ and is therefore not a rational number. This number can, however, be associated with a point on the number line. In fact, by applying the Pythagorean Theorem, the square roots of all the whole numbers can be represented geometrically as line segments and associated with points on the number line as shown in Figure 5.11.1. Some of these numbers are rational and some are not; for example, $\sqrt{4}$, $\sqrt{9}$, and $\sqrt{16}$ are rational whereas $\sqrt{2}$, $\sqrt{3}$, and $\sqrt{5}$ are not.

We have just said that $\sqrt{2}$ cannot be represented as an ordered pair of integers $\frac{a}{b}$ and hence is an irrational number. To demonstrate this claim

FIGURE 5.11.1

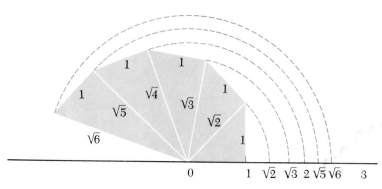

we will use an *indirect proof* (*proof by contradiction*). In this type of proof we assume as true the *negation* of the statement we wish to prove and, by the implications of this assumption, arrive at a conclusion known to be false. Since our mathematical procedures were correct, it follows that we arrived at a contradiction because our assumption was false. Hence the original statement, rather than its negation, must be true. (Refer to Section 1.8 for further details concerning the nature of an indirect proof.)

Proof that $\sqrt{2}$ is an irrational number: Assume that $\sqrt{2}$ is a rational number and that $\sqrt{2} = \frac{a}{b}$, where $\frac{a}{b}$ is a fraction in simplest form; that is, a and b have no common positive integral factor other than 1. Then $2 = \frac{a^2}{b^2}$ and $2b^2 = a^2$. Since $2b^2$ has a factor of 2, a^2 must have a factor of 2. However, $a^2 = a \cdot a$, and each factor a must have a factor of 2, since 2 is prime and a number can be factored into primes in only one way. (The fact that a number can be factored into primes in only one way is discussed in Section 6.3.) Since a has a factor of 2, let $a = 2k$. Then by substitution, $2b^2 = (2k)^2$ and $2b^2 = 4k^2$. Thus $b^2 = 2k^2$. By again reasoning as above, b^2 must have a factor of 2 and hence b has a factor of 2. It has now been shown that a and b have a common factor of 2. Hence our assumption that $\frac{a}{b}$ is a fraction in simplest form is contradicted, and this inconsistency leads us to conclude that $\sqrt{2}$ is not a rational number because any rational number can be represented by a fraction in simplest form. Therefore, $\sqrt{2}$ must be an irrational number.

By proofs similar to the above, it can also be shown that numbers such as $\sqrt{3}$, $\sqrt{5}$, and $\sqrt{13}$ are irrational. Other examples of irrational numbers are $\sqrt[5]{7}$, $\log_{10} 5$, and π. There are many irrational numbers. In fact, it can be proved that there are more irrational numbers than rational numbers, even though there are infinitely many elements in both sets. Both the rationals and the irrationals are dense sets.

Many interesting stories can be told concerning the discovery and study of irrational numbers. Pythagoras and his select group of followers, known as the Pythagoreans, originally believed that all numbers were rational. Furthermore, the Pythagoreans were a semireligious fraternity, and had committed themselves to the concept that the entire universe could be explained in terms of the counting numbers. Hence they became quite disturbed when it was discovered that $\sqrt{2}$ could be used to represent a distance in the universe but could not be written as $\frac{a}{b}$ where a and b are counting numbers. To avoid embarrassment, the Pythagoreans decided to keep their knowledge of irrational numbers a secret; but the inevitable happened and one member, Hippasus, revealed the discovery. According to legend, Hippasus was lost a short time later in a tragic "accident" while sailing with other members of the Pythagorean brotherhood. Did he fall or was he pushed? We will probably never know.

Since there are infinitely many rational numbers between any two given rationals, and since these numbers may be associated with infinitely many points between two points on the number line, it may at first seem intuitively evident that every point on the number line will be associated with some rational number; but as we have seen, our intuition fails, for while every rational number may be associated with exactly one point on the number line, infinitely many points on the number line are *not* associated with any rational number. However, by inventing the *irrationals*, we can now "fill in the gaps" between the rationals, and each point on the number line can then be associated uniquely with either a rational number or with an irrational number; more will be said about this in Section 5.12.

EXERCISE 5.11

1. Refer to Figure 5.10.2 and prove the Pythagorean Theorem. You may wish to use more steps in your proof than are given in the text.

2. Using the figure below, prove the Pythagorean Theorem by showing that $a^2 + b^2 = c^2$. [*Hint:* Use the relation $x = b - a$.]

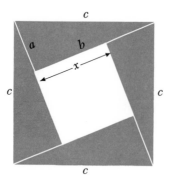

3. (a) If the hypotenuse of a right triangle measures 13 feet and one of the legs measures 5 feet, what is the length of the other leg?
   (b) If the legs of a right triangle measure 2 units and 3 units, what is the length of the hypotenuse?

4. What is the length of the side of a square inscribed in a circle with a radius of 5 units?

5. Prove that $\sqrt{2}$ is irrational (without referring to the text unless necessary).

6. Prove that $\sqrt{3}$ is irrational by a method similar to that used in Problem 5. [*Hint:* A number with a factor of 3 may be written in the form $3k$.]

7. (a) Assume that $\sqrt{2}$ is known to be irrational and prove that $7 + \sqrt{2}$ is irrational. [*Hint:* Assume that $7 + \sqrt{2}$ is a rational number and use an indirect proof.]

   (b) Prove that $8 - \sqrt{3}$ is irrational if it is known that $\sqrt{3}$ is irrational.

8. (a) Find the sum $(8 - \sqrt{3}) + (8 + \sqrt{3})$. Is the sum irrational?

   (b) Find the difference $(8 - \sqrt{3}) - (8 + \sqrt{3})$. Is the difference irrational?

   (c) Find the product $(8 - \sqrt{3})(8 + \sqrt{3})$. Is the product irrational?

5.12 The Real Numbers

The set that results from the union of the set of rational numbers and the set of irrational numbers is called the set of **real numbers**. The set of real numbers may be put in a one-to-one correspondence with the set of *all* points on the number line; that is, not only can every real number be associated with a unique point on the number line, but every point on the number line can also be associated with a unique real number. Thus the number line may be used as a picture of the set of real numbers in such a way that there are no "holes" in the ordering of the real numbers. The technical way of expressing this concept is to say that the real numbers are **order-complete**. While the rationals and the irrationals are *dense* sets, neither set is *order-complete*.

Since we can identify any point on a line with a real number, we can use the real numbers to engage in such activities as measuring the length of *any* given line segment. By extending the geometric application of the real numbers, any point in a plane can be identified with an ordered pair of real numbers, and any point in space can be identified with an ordered triplet of real numbers. Furthermore, we can construct a **system of real numbers** with all of the properties for addition, multiplication, and so on, that the rational numbers have. Every property of the rationals as stated in Definition 5.8.1 holds for the real number system.

Although the set of real numbers, which is comprised of the rationals and the irrationals, has many advantages, computing with either the rationals or with the irrationals can be difficult and tedious. Consider, for example, the problem encountered in finding the sum $\frac{237}{428} + \frac{3256}{8271}$ or perhaps the quotient $642 \div \sqrt{947}$.

As you know, computing in the integers is relatively easy if decimal numerals are used. (But not so easy if Roman numerals are used.) Fortunately, the decimal system of numeration can be extended so that all real numbers, both rational and irrational, can be written in decimal form. Every *rational* number may be indicated either by a *terminating decimal* such as 2.49, or by a *repeating infinite decimal* such as 0.373737. ... . Every *irrational* number may be indicated by a *nonrepeating infinite decimal* such as 3.141592. ... . First let us consider the decimal representation of the rational numbers.

5.13   The Decimal Representation of Rational Numbers

Simon Stevin (1548–1620) of the Netherlands is credited with developing some of the initial concepts in the use of *decimal fractions*. He was no doubt amply motivated to develop simpler numerals while serving as inspector of the dikes, quartermaster-general of the Dutch army, and minister of finance. In addition to making a genuine contribution to the field of mathematics, Stevin also wrote on physics and hydraulics.

    Of importance to us at this time is Stevin's systematic treatment of decimal fractions in which he showed how they could be used in the place of ordinary fractions. In 1608 Robert Norton published an English translation of Stevin's work. Norton was so impressed with the value of decimals that he offered as a public service free instruction on their use.

    Stevin's method for extending the decimal system to include the fractions is accomplished in modern notation by placing a dot, called a **decimal point**, after the units digit of a numeral and letting the digits to the right of the decimal point represent an integral number of tenths, hundredths, thousandths, and so on. If there is no units digit in a given numeral, a zero is sometimes placed to the left of the decimal point to make the decimal point conspicuous. Examples are 0.35 and 0.0042.

    A fraction may be expressed as a **terminating decimal** only if it is possible to write the fraction with a denominator that is an integral power of 10. Since $10 = 2 \cdot 5$, $10^2 = 2^2 \cdot 5^2$, $10^3 = 2^3 \cdot 5^3$, or in general, $10^n = 2^n \cdot 5^n$, where n is an integer, we see that the prime factorization of the denominator may contain only integral powers of 2 and 5. For example, $\frac{3}{40}$ may be expressed as a terminating decimal since its denominator $40 = 2 \cdot 2 \cdot 2 \cdot 5$ and we have:

$$\frac{3}{40} = \frac{3}{2 \cdot 2 \cdot 2 \cdot 5} = \frac{3}{2^3 \cdot 5} = \frac{3(5^2)}{2^3 \cdot 5(5^2)} = \frac{3 \cdot 25}{2^3 \cdot 5^3} = \frac{3 \cdot 25}{(2 \cdot 5)^3}$$
$$= \frac{75}{10^3} = \frac{75}{1000} = 0.075.$$

(Of course, we could have obtained the same result by the familiar process of dividing numerator by denominator.)

    It is easy to see from the above example that if the denominator of a fraction has more factors of 2 than factors of 5, or more factors of 5 than factors of 2, we can write an equivalent fraction so that the denominator will have an equal number of factors of 2 and 5 and thus will equal an integral power of 10.

At first, one might believe that a fraction such as $\frac{3}{6}$ cannot be written with a denominator that is an integral power of 10, since $6 = 2 \cdot 3$ and 3 is not a prime factor of 10. However, since the fraction $\frac{3}{6}$ is not in simplest form, we see that $\frac{3}{6} = \frac{1}{2} = \frac{1}{2} \cdot \frac{5}{5} = \frac{5}{10} = 0.5$. It is easier to determine whether a fraction may be written as a terminating decimal if the fraction is first written in simplest form. In fact, a general rule is that if a fraction is written in simplest form and if its denominator contains prime factors of only 2's and 5's, then it may be expressed as a terminating decimal. Fractions such as $\frac{5}{6}$, $\frac{1}{7}$, and $\frac{4}{21}$ cannot be expressed as terminating decimals. Note that these fractions are in simplest form and have denominators with prime factors other than 2 and 5.

If a rational number given as a fraction in simplest form has a denominator with prime factors other than 2's and 5's, the decimal representation may be found by dividing the numerator of the fraction by the denominator; in this case the result is a **repeating infinite decimal**. By a *repeating decimal* we mean a numeral in which a particular block of consecutive digits, called the **repetend**, is repeated again and again in a nonterminating sequence. For example, $0.161616\ldots$ is a repeating decimal whose repetend is 16.

Below we give a few examples of rational numbers expressed first as fractions and then as repeating decimals. In each case the repeating decimal is found by dividing the numerator of the fraction by the denominator. The notation is simplified by placing a bar over the repetend.

Example 1.
(a) $\frac{13}{30} = 0.4333\ldots = 0.4\overline{3}$
(b) $\frac{2}{3} = 0.666\ldots = 0.\overline{6}$
(c) $\frac{4}{11} = 0.363636\ldots = 0.\overline{36}$
(d) $\frac{5}{7} = 0.714285714285\ldots = 0.\overline{714285}$

We emphasize that not all the digits in a repeating decimal need repeat. In Example 1(a), for instance, the bar is placed only over the digit 3, indicating that only this digit repeats; the digit 4 does not repeat.

What is the maximum number of digits possible in the repetend of a repeating decimal? As an aid in answering this question, let us consider the number $\frac{5}{7}$ and examine the division algorithm used in finding its repeating decimal.

Example 2.

$$
\begin{array}{r}
0.7\ 1\ 4\ 2\ 8\ 5 \\
7\overline{)5.0\ 0\ 0\ 0\ 0\ 0} \\
4\ 9 \\
\overline{①\ 0} \\
7 \\
\overline{③\ 0} \\
2\ 8 \\
\overline{②\ 0} \\
1\ 4 \\
\overline{⑥\ 0} \\
5\ 6 \\
\overline{④\ 0} \\
3\ 5 \\
\overline{⑤}
\end{array}
$$

In the division algorithm of Example 2, the remainders have been circled. Since any remainder must be less than the divisor 7, the only possible remainders are 0, 1, 2, 3, 4, 5, and 6. However, if a remainder of 0 is obtained, the decimal terminates. Therefore, if dividing by 7 produces a repeating decimal, there are only six possible remainders: 1, 2, 3, 4, 5, and 6. This means that there will be a *maximum* of six digits in the repetend of the decimal representation of a fraction with a denominator of 7. In the case of $\frac{5}{7}$, the maximum is reached since all possible nonzero remainders are obtained in the division algorithm.

In determining the maximum number of digits possible in the repetend of a repeating decimal, it is important to examine the simplest fraction form of the rational number that it represents. Since $\frac{10}{14} = \frac{5}{7}$, for example, $\frac{10}{14}$ can have no more digits in the repetend of its decimal representation than $\frac{5}{7}$ has.

In general, if the *simplest* fraction form $\frac{a}{b}$ of a rational number is written as a repeating decimal, there will be a *maximum* of $b - 1$ digits in the repetend of the decimal. Of course, many decimals repeat before the maximum possible number of digits is reached. For example, although the decimal representation of $\frac{4}{11}$ has a possible maximum of $11 - 1$, or 10, repeating digits, $\frac{4}{11}$ is equal to $0.\overline{36}$, which has only 2 repeating digits. Some other number with a denominator such as 3821 may actually have 3820 digits in the repetend of its decimal representation; however, we can be assured that it will have no more than that number. (This may be only slight comfort to us, but it is very helpful for a computer.)

We have seen that it is possible to write every rational number either as a terminating or as a repeating decimal by using the division process of Example 2. Conversely, it is also possible to express every terminating or repeating decimal as a fraction of the form $\frac{a}{b}$. The procedure for finding the fraction form of a rational number when it is expressed as a repeating infinite decimal is shown by Example 3.

Example 3. Find the fraction form of the rational number represented by $0.\overline{36}$.

Solution: Let $x = 0.\overline{36}$.

Then
$$100x = 36.3636\ldots$$
Subtract
$$x = 0.3636\ldots$$
$$99x = 36$$
$$x = \frac{36}{99} = \frac{4}{11}$$

Therefore, $0.\overline{36} = \frac{4}{11}$.

In Example 3 we multiply each member of the equation $x = 0.\overline{36}$ by 10^2 or 100 because there are 2 digits in the repetend. If there were n repeating digits, we would multiply each member of the equation by 10^n. The reason for this is apparent as soon as we get to the step where the operation of subtraction is performed. Another example should be sufficient to illustrate the process.

Example 4. Find the fraction form of the rational number represented by $3.4\overline{263}$.

Solution: Let $x = 3.4\overline{263}$. We now multiply each member of this equation by 10^3 or 1000 because there are 3 digits in the repetend.

$$1000x = 3426.\overline{3263}$$
$$x = 3.4\overline{263}$$
$$999x = 3422.9$$
$$x = \frac{3422.9}{999} = \frac{34,229}{9990}$$

Therefore, $3.4\overline{263} = \frac{34,229}{9990}$.

Before we leave the topic of repeating decimals, let us mention that every terminating decimal can be written as a repeating infinite decimal by subtracting 1 from its last digit and annexing an infinite number of 9's. For example, $0.27 = 0.26\overline{9}$, $0.4 = 0.3\overline{9}$, $19.32 = 19.31\overline{9}$, and $4 = 3.\overline{9}$.

Example 5. Prove $0.26\overline{9} = 0.27$.

Solution: Let $x = 0.26\overline{9}$.

Then $\qquad\qquad 10x = 2.699\overline{9}$

Subtract $\qquad\qquad\quad x = 0.269\overline{9}$

$$9x = 2.43$$

$$x = \tfrac{2.43}{9} = 0.27.$$

Therefore, $0.269\overline{9} = 0.27$.

In order to have a unique decimal representation for each rational number, we do not ordinarily represent terminating decimals as repeating infinite decimals. It is nevertheless somewhat intriguing to find that this can be done.

5.14 The Decimal Representation of Irrational Numbers

While each rational number may be represented by a terminating decimal or by a repeating infinite decimal, each irrational number may be represented by a nonrepeating infinite decimal. Examples of irrational numbers are $\sqrt{2} = 1.41424\ldots$, $\sqrt{3} = 1.73205\ldots$, $\pi = 3.141592\ldots$, $\log 7 = 0.8451\ldots$, and $\sin 45° = 0.7071\ldots$.

Since it is obvious that we cannot write infinitely many digits, we frequently *approximate* an irrational number by using only a finite number of digits. Such an approximation is, of course, a *rational* approximation, since a terminated decimal may be written as an ordered pair of integers $\frac{a}{b}$. A few examples of rational approximations of irrational numbers appear in Example 1. Note that the symbol $\approx$ is used for "is approximately equal to."

Example 1.

$$\sqrt{2} = 1.41421\ldots \approx 1.414$$
$$\sqrt{3} = 1.73205\ldots \approx 1.732$$
$$\pi = 3.14159\ldots \approx 3.1416$$
$$\log 7 = 0.8451\ldots \approx 0.85$$
$$\sin 45° = 0.7071\ldots \approx 0.7$$

It is natural to wonder how the digits of the decimal representation of an irrational number can be determined. The way in which an irrational number is defined frequently gives us a hint. The $\sqrt{2}$, for example, is by definition the number that used as a factor twice is equal to 2. Hence we can find the initial digits in the decimal representation of $\sqrt{2}$ simply by "trial and error." The first digit must be 1 since $1^2 = 1$, which is less than 2, while $2^2 = 4$, which exceeds the number 2. Similarly, the next digit must be 4 since $(1.4)^2 = 1.96$, which is less than 2, while $(1.5)^2 = 2.25$, which exceeds 2. Continuing this line of reasoning, we can determine, as shown below, more and more of the digits in the decimal numeral for $\sqrt{2}$.

$$1^2 = 1$$
$$(1.4)^2 = 1.96$$
$$(1.41)^2 = 1.9881$$
$$(1.414)^2 = 1.999396$$
$$(1.4142)^2 = 1.99996164$$
$$(1.41421)^2 = 1.9999899241$$

For many irrational numbers, it is difficult, time consuming, and tedious to determine the digits of the decimal representation of the number. Sometimes, however, we can find an infinite series that is equal to the given irrational number. For example, $\pi = \frac{4}{1} - \frac{4}{3} + \frac{4}{5} - \frac{4}{7} + \frac{4}{9} - \frac{4}{11} + \cdots$. Adding more and more terms of this series (each of which may be expressed in decimal form), results in a progressively better approximation of π. Frequently a knowledge of calculus is necessary, or at least desirable, to determine the series for a given number.

Other irrational numbers involving π are

$$\frac{\pi^2}{6} = \frac{1}{1^2} + \frac{1}{2^2} + \frac{1}{3^2} + \cdots$$
$$\frac{\pi^2}{8} = \frac{1}{1^2} + \frac{1}{3^2} + \frac{1}{5^2} + \cdots$$

The value of an automatic electronic computer in evaluating series, such as the preceding, is obvious.

Some irrational numbers may be expressed as the product of infinitely many factors; for example:

$$\frac{\pi}{2} = \left(\frac{2\cdot2}{1\cdot3}\right)\left(\frac{4\cdot4}{3\cdot5}\right)\left(\frac{6\cdot6}{5\cdot7}\right)\left(\frac{8\cdot8}{7\cdot9}\right) \cdots$$

The final result, of course, may be expressed in decimal form, and by using more and more factors, a better and better decimal approximation of π can be obtained.

For some people, the study of infinite series and various sequences is one of the most interesting and exciting topics in mathematics. The more one learns, the more interesting things become. It seems that for each question answered, others arise to take its place.

EXERCISE 5.14

1. Find the terminating or repeating decimal for each of the following and indicate repeating digits by placing a bar over them.

(a) $\frac{7}{33}$ (b) $\frac{2}{3}$ (c) $\frac{1}{8}$

(d) $\frac{7}{8}$ (e) $\frac{10}{33}$ (f) $\frac{3}{11}$

(g) $\frac{5}{7}$ (h) $\frac{9}{11}$ (i) $\frac{28}{111}$

2. Find a fraction numeral for the rational number represented by each of the following repeating decimals.
   (a) $0.\overline{4}$
   (b) $0.6\overline{3}$
   (c) $0.\overline{54}$
   (d) $0.\overline{90}$
   (e) $0.3\overline{93}$
   (f) $0.4\overline{23}$
   (g) $98.\overline{7}$
   (h) $42.3\overline{54}$
   (i) $0.\overline{101}$

3. (a) How many nonzero remainders are possible when a whole number is divided by 13?
   (b) If a whole number is divided by 13, what is the maximum possible number of digits in the repetend of the quotient?
   (c) Find the repeating decimal for $\frac{5}{13}$.

4. The maximum possible number of digits in the repetend of the quotient $20 \div 52$ is 12. Why is this true?

5. Show that:
   (a) $0.44\overline{9} = 0.45$
   (b) $1.\overline{9} = 2$
   (c) $0.\overline{9} = 1$
   (d) $7.23\overline{9} = 7.24$

6. If a, b, c, and d are the digits of the repeating decimal $0.\overline{abcd}$, find a fraction to represent this number.

7. By "trial and error" find the first two digits of the decimal numeral for each of the following irrational numbers:
   (a) $\sqrt{7}$
   (b) $\sqrt{39}$

5.15   Imaginary, Complex, and Transfinite Numbers

In a single equation such as $3x = 12$, it is obvious that the only solution is $x = 4$, since $3(4) = 12$. On the other hand, for the equation $x^2 = 9$, we see that there are two solutions, $+3$ and -3. This is true since $(+3)^2 = (+3)(+3) = 9$ and $(-3)^2 = (-3)(-3) = 9$. Let us reconsider the equation $x^2 = 9$. By definition, x is the square root of 9; but there are two square roots of 9, one positive and one negative. Hence we can write $x = +\sqrt{9} = +3$, or $x = -\sqrt{9} = -3$. These two results may be written together as $x = \pm\sqrt{9} = \pm3$. Similarly:

If $x^2 = 4$, then $x = +\sqrt{4} = +2$, or $x = -\sqrt{4} = -2$.
If $x^2 = 3$, then $x = +\sqrt{3}$, or $x = -\sqrt{3}$.
If $x^2 = 2$, then $x = +\sqrt{2}$, or $x = -\sqrt{2}$.
If $x^2 = 1$, then $x = +\sqrt{1} = +1$, or $x = -\sqrt{1} = -1$.

Some of the above equations have rational solutions while others have irrational solutions, but *all* have *real number solutions*. Now we are prepared to consider something slightly different.

Suppose $x^2 = -1$. We cannot find any *real* number solution for this equation, since the square of every real number is either a positive number or zero and hence cannot equal -1. Analogous with the above, however, we are tempted to write $x = +\sqrt{-1}$ or $x = -\sqrt{-1}$. Thus, we invent numerals for two new numbers which are the two square roots of -1. If we wish to abbreviate our symbolism even further, we may write $+\sqrt{-1} = i$ and $-\sqrt{-1} = -i$. Similarly:

If $x^2 = -2$, then $x = +\sqrt{-2} = +\sqrt{(2)(-1)} = +\sqrt{2}\sqrt{-1} = +\sqrt{2}i$, or $x = -\sqrt{-2} = -\sqrt{(2)(-1)} = -\sqrt{2}\sqrt{-1} = -\sqrt{2}i$.

If $x^2 = -3$, then $x = +\sqrt{3}i$, or $x = -\sqrt{3}i$ (leaving out the intermediate steps).

If $x^2 = -4$, then $x = +\sqrt{4}i = +2i$, or $x = -\sqrt{4}i = -2i$.

If $x^2 = -5$, then $x = +\sqrt{5}i$ or $x = -\sqrt{5}i$

If $x^2 = -\pi$, then $x = +\sqrt{\pi}i$, or $x = -\sqrt{\pi}i$

These new numbers are (unfortunately) called imaginary numbers. Undoubtedly, they were so named because initially they were not well understood. In 1702 the great German mathematician Gottfried Leibniz, who shares with Isaac Newton credit for developing calculus, wrote, "The imaginary numbers are a wonderful flight of God's Spirit; they are almost an amphibian between being and not being." Even though the imaginary numbers are just as "real" as the real numbers, the names imaginary and real have remained in our vocabulary; and it seems little will be done about it.

In general, the imaginary numbers are written in the form bi, where b is a real number and $i = \sqrt{-1}$. Hence, $17i$, $-8i$, $\sqrt{19}i$, πi, and $\frac{2}{7}i$ are all examples of imaginary numbers.

The imaginary numbers may be placed in a one-to-one correspondence with the points on a number line in a manner similar to the one used for the real numbers. If, for the imaginary numbers, we choose a line perpendicular to the real number line, the result is as shown in Figure 5.15.1.

If a real number is added to an imaginary number we have a number whose numeral may be written in the form $a + bi$, where a and b are real numbers and once again $i = \sqrt{-1}$. A number of the form $a + bi$ is called a complex number. A complex number may also be represented simply as an ordered pair of real numbers (a, b). Examples of complex numbers are $2 + 3i$, $2 + (-3)i = 2 - 3i$, $\frac{1}{2} - \sqrt{2}i$, and $\pi + \sqrt{7}i$.

Observe that if $a = 0$, then the complex number $a + bi = 0 + bi = bi$, which is an imaginary number (sometimes called pure imaginary). Similarly, if $b = 0$, then $a + bi = a + 0i = a + 0 = a$, which is a real number. Hence the real numbers and the pure imaginary numbers are proper subsets of the complex numbers.

The complex numbers may be placed in a one-to-one correspondence

FIGURE 5.15.1

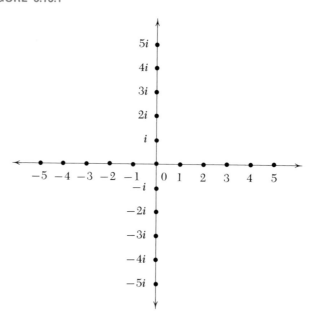

with all of the points in a plane. Figure 5.15.2 illustrates this fact by using a few representative numbers and points. All of the real numbers are on the horizontal line, called the *real axis*, while the pure imaginary numbers are on the vertical axis, called the *imaginary axis*. The rest of the complex numbers are on neither axis.

The German mathematician Carl Frederick Gauss (1777–1855) was one of the first to present complex numbers in a meaningful way and to interpret them as labeling the points in a plane. Although Gauss was truly a genius, and ranks with the very best of mathematicians by any standard, he did not bring about the acceptance of complex numbers single-handed. While Gauss introduced the term "complex number" in 1832, René Descartes used the terms "real" and "imaginary" as early as 1748. Acceptance of the complex numbers was slow, and even in the late 1800's some mathematicians were uncertain about the value of these new numbers. After practical applications for complex numbers were discovered by men such as Charles P. Steinmetz (1865–1923), who used them in developing the theory of electrical circuits, they became accepted by nearly everyone as "legitimate" numbers. The Venn diagram of Figure 5.15.3 illustrates how the various sets of numbers discussed in this chapter are related to each other.

There is at least one significant difference between the complex numbers and the real numbers; complex numbers cannot be ordered according to size as can the real numbers. We cannot say, for example, that

FIGURE 5.15.2

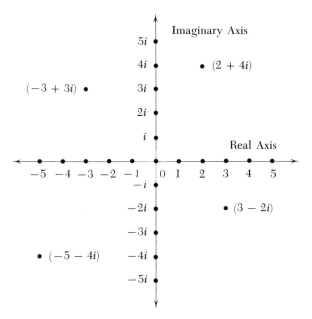

$8 + 3i$ is less than or greater than $2 + 9i$, in the same sense that we can say 6 is less than 7.

We will not formally define addition and multiplication for the complex numbers. However, sums and products are exactly what you might expect them to be as the next example shows.

Example 1.
(a) $(2 + 3i) + (5 + 9i) = 7 + 12i$
(b) $(4 + 5i) + (6 - 2i) = 10 + 3i$
(c) $(-2 + 3i) + (9 - 8i) = 7 - 5i$
(d) $i^2 = (\sqrt{-1})^2 = (\sqrt{-1})(\sqrt{-1}) = -1$
(e) $(2 + 3i)(6 + 5i) = 12 + 10i + 18i + 15i^2$
$\qquad\qquad\quad = 12 + 28i + 15(-1)$
$\qquad\qquad\quad = 12 + 28i - 15$
$\qquad\qquad\quad = -3 + 28i$

Although we will not give the proofs here, it is relatively easy to show that the system of complex numbers has all of the properties of the rational numbers (and real numbers) that are stated in Definition 5.8.1.

Are there still other numbers that are not included in the set of complex numbers? The answer is yes. For example, special numbers have been invented to tell us how many elements are in certain sets with infinitely

FIGURE 5.15.3

Sets of Numbers with Examples of Each

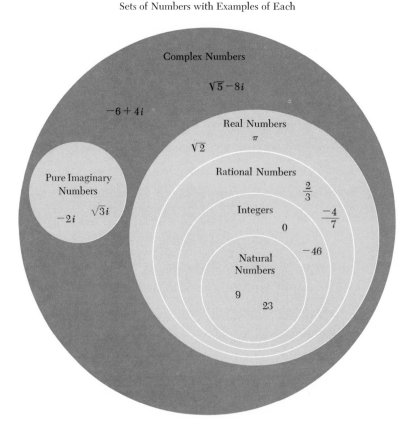

many elements. Georg Cantor, referred to in Chapter 2 for his work in set theory, did much to establish theories to deal with cardinal numbers for infinite sets; such numbers are called **transfinite numbers**. His work on transfinite numbers was developed and published in a sequence of papers between the years 1870 and 1895.

As you know, two finite sets have the same cardinal number if their elements may be placed in a one-to-one correspondence. Essentially, Cantor extended this concept of cardinality to include infinite sets. Thus, it turns out that the cardinal number of the set of even natural numbers is the same as the cardinal number of the entire set of natural numbers. If this does not seem correct to you, look at the following method that is used to place the elements of the two sets in a one-to-one correspondence.

$$
\begin{array}{cccccccc}
1 & 2 & 3 & 4 & 5 & & n & \\
\updownarrow & \updownarrow & \updownarrow & \updownarrow & \updownarrow & \cdots & \updownarrow & \cdots \\
2 & 4 & 6 & 8 & 10 & & (2n) &
\end{array}
$$

The cardinal number of the set of natural numbers is named $\aleph_0$, read "aleph null" ($\aleph$ is the first letter of the Hebrew alphabet). Hence the transfinite cardinal number of any other set (such as the set of even numbers) whose elements may be placed in a one-to-one correspondence with the natural numbers also has the transfinite cardinal number $\aleph_0$. A remarkable result of this theory is that a proper subset of an infinite set may have the same cardinal number as the set itself. This, of course, is *not* true for finite sets. In fact, it is this property that characterizes infinite sets as opposed to finite sets.

Example 2. Show that $4 + \aleph_2 = \aleph_0$.

Solution: We know that $4 = n(\{a, b, c, d\})$ and $\aleph_0 = n(N)$. By definition of addition, $4 + \aleph_0 = n(\{a, b, c, d\} \cup N)$. However, we can place the elements of $\{a, b, c, d\} \cup N$ in a one-to-one correspondence with the elements of N as follows:

$$\{a, b, c, d\} \cup N = \{a, b, c, d, 1, 2, 3, 4, \ldots, \quad n, \quad \ldots\}$$
$$N = \{1, 2, 3, 4, 5, 6, 7, 8, \ldots, (n + 4), \ldots\}$$

Hence the two sets have the same cardinal number and we conclude that $4 + \aleph_0 = \aleph_0$.

Although we will not show the proofs here, both the integers and the rational numbers may be placed in a one-to-one correspondence with the natural numbers; hence each of these sets has the cardinal number $\aleph_0$. The real numbers, however, *cannot* be placed in a one-to-one correspondence with the natural numbers and have a larger infinite cardinal number, which we may name by using $\aleph_1$, or some other symbol.

The theories of infinite sets and transfinite numbers lead to some paradoxes which remain (as of this writing) unsolved. For example, it can be proved that there is a cardinal number greater than the largest possible cardinal number. Regardless of the paradoxes and unsolved problems concerning sets and transfinite numbers, the works of Cantor will remain a landmark in mathematics. At times man's creativity seems almost boundless, while at other times his abilities seem very limited. He has flown to the moon, but cannot resolve seemingly simple paradoxes. Perhaps the paradoxes of today will eventually be resolved.

EXERCISE 5.15

1. If $i = \sqrt{-1}$, find the equivalent numerals for:
   (a) i^2 (b) i^3 (c) i^4 (d) i^5 (e) i^6
   (f) Do you see a pattern emerging?

2. Solve each of the following equations.
   (a) $x^2 = 49$
   (b) $x^2 = 81$
   (c) $x^2 = -49$
   (d) $x^2 = -81$
   (e) $x^2 = -144$

3. To Isaac Newton is attributed the saying "If I have seen a little farther than others it is because I have stood on the shoulders of giants."
   (a) What is the meaning of this metaphoric statement?
   (b) How does this statement apply to the contributions of Gauss in developing the complex numbers?

4. Find each of the following sums and products and write the result in the form $a + bi$.
   (a) $(2 + 3i) + (7 + 4i)$ (b) $(2 - 4i) + (7 + 6i)$
   (c) $(-2 + 3i) + (4 - 4i)$ (d) $(3 + 4i)(6 + 2i)$
   (e) $(3 + 4i)(2 - 3i)$ (f) $(-2 + 3i)(2 + 2i)$

5. Consider the set whose elements are the squares of the integers: $\{1, 4, 9, 16, 25, \ldots\}$. Show that this set has the same cardinal number as the set of natural numbers.

6. Show that the set of odd numbers has the cardinal number $\aleph_0$.

7. Prove that $3 + \aleph_0 = \aleph_0$.

8. Using sets show that $\aleph_0 - 5 = \aleph_0$.

9. Devise a method to prove that $\aleph_0 + \aleph_0 = \aleph_0$.

REVIEW EXERCISES

1. What integer is equal to n if:
   (a) $-17 + n = 0$ (b) $23 + n = 0$
   (c) $n + 4 = 1$ (d) $7 + n = -9$

2. Show by a counterexample that subtraction in the integers is not commutative.

3. Using the fact that $a - b = a + (-b)$, find the following differences:
   (a) $8 - 12$ (b) $6 - (-4)$
   (c) $-9 - 6$ (d) $0 - (-15)$

4. Using the definition for the sum of two rational numbers, find the following sums and express each result in simplest form:
   (a) $\frac{3}{8} + \frac{7}{16}$ (b) $\frac{-2}{3} + \frac{4}{9}$

5. Use *addition* in finding the following differences:
   (a) $\frac{2}{3} - \frac{4}{5}$ (b) $\frac{6}{6} - \frac{-3}{4}$

6. Perform the indicated operations:
   (a) $\frac{3}{5} \div \frac{-4}{3}$ (b) $\frac{2}{3} \div -5$
   (c) $0 \div \frac{2}{3}$ (d) $-6 \div \frac{2}{3}$

7. Insert the correct symbol ($<$, $=$, or $>$) between the numbers in each of the following:

    (a) $\frac{7}{11}$, $\frac{5}{8}$ (b) $\frac{11}{10}$, $\frac{10}{9}$

    (c) -6, -5 (d) $\frac{-2}{3}$, $\frac{-1}{5}$

8. Find the repeating decimal equivalent to:

    (a) $\frac{4}{15}$ (b) $\frac{4}{7}$

9. Find a fraction numeral for the rational number represented by each of the following repeating decimals:

    (a) $0.\overline{23}$ (b) $0.44\overline{9}$

10. For each of the following perform the indicated operation and write the result in the form of a complex number $a + bi$.

    (a) $(3 + 5i) + (8 + i)$ (b) $(2 - 3i) + (7 + 2i)$

    (c) $(4 + 2i)(3 + 5i)$ (d) $(-3 + 4i)(6 + 5i)$

Vasily Kandinsky, *Extended, No. 333*
1926. Oil on wood. 37½ x 17½".
The Solomon R. Guggenheim Museum, New York

THE THEORY
OF NUMBERS

6.1 Introduction

For centuries, the study of the properties of the integers and their relationships with each other has fascinated man. Such studies resulted in the belief that certain numbers had great mystical significance. Even today many public buildings do not have a floor numbered 13! Do you know people who consider 7 a lucky number? Every year, the author's wife receives a check for $7 from her father upon celebrating her birthday. (Because of inflation, this has become $17.)

The study of the integers and their properties is known as the theory of numbers. Historical records show that the Chinese and the Egyptians were interested in number theory as far back as 2000–3000 years B.C. More recently, the Frenchman Pierre de Fermat (1601–1665), who has been referred to as the father of modern number theory, helped establish the theory of numbers as a recognized branch of mathematics. Fermat was a lawyer and civil servant, but his favorite pastime was mathematics. Although Fermat is credited with inventing analytic geometry independently of René Descartes, since he and Descartes frequently communicated by letter, each probably learned from the other. In any event, Fermat made many contributions to mathematics, but is best known for his work in number theory concerning the properties of the primes. We shall discuss a special theorem named after Fermat in Section 6.5.

Perhaps the most prolific contributor to the theory of numbers was Carl Friedrich Gauss, whom we mentioned previously for his work with complex numbers. In fact, Gauss made advances in many branches of mathematics as well as astronomy, geodesy (measuring and studying the shape of the earth), and electromagnetism. While the names of Cooke, Wheatstone, and Morse are usually associated with making the first electric telegraph systems, Gauss apparently predated them by inventing an electric telegraph in 1833. He used it to send messages to a fellow worker, Wilhelm Weber (1804–1891). Although Gauss seemed to care little about practical applications, Weber predicted in 1835 that "when the globe is covered with a net of railroads and telegraph wires, this net will render services comparable to those of the nervous system in the human body, partly as a means of transport, partly as a means for the propagation of ideas and sensations with the speed of lightning."

Gauss' many contributions to mathematics earned him the title "The Prince of Mathematicians." He was certainly not called a prince because of his noble birth, since his parents were people of very modest means. His father was a gardener and a bricklayer, and his mother was the daughter of a stonecutter. Gauss learned to read and compute in arithmetic before the age of three. Although his father did not encourage Gauss to engage in intellectual pursuits, his mother, a teacher named Büttner, and a young assistant schoolmaster named Johann Martin Bartels gave all the support that was necessary. Gauss attended several schools including the University of Göttingen and produced much more mathematics before the age of 21 than anyone would expect in a lifetime.

Unfortunately, Gauss cared little for publicity and many of his mathematical discoveries were not published. Other mathematicians spent hours rediscovering things already known to Gauss. We know this to be true because Gauss kept a scientific diary from the age of twenty. The diary first came into the scientific community in 1898, 43 years after the death of Gauss; it was made available for study when the Royal Society of Göttingen borrowed it from a grandson of Gauss.

Although many interesting pages could be written on the life of Gauss, our primary concern at this time is, of course, the fact that he made so many contributions in the theory of numbers. Number theory is one of the few branches of mathematics in which the layman, as well as the professional mathematician, has been able to make unique contributions.

Even though it is becoming more difficult to develop original results without specialized training, still number theory is a field where anyone may have the satisfaction of rediscovering certain remarkable properties and relationships. In this chapter we shall very briefly introduce a few of the many topics in the theory of numbers.

6.2 Prime and Composite Numbers

In studying the whole numbers, we mentioned that a and b are called *factors* of the *product* $a \cdot b$; the product $a \cdot b$ is also said to be a *multiple* of both a and b. Some whole numbers have many distinct factors while others have few. For example, the factors of 20 comprise the set $F_{20} = \{1, 2, 4, 5, 10, 20\}$, while the factors of 47 are the members of the set $F_{47} = \{1, 47\}$. A number such as 47 which has *only* itself and 1 as factors is said to be *prime*. Every whole number, of course, has itself and 1 as factors since 1 is the multiplicative identity and $n = n \cdot 1$.

DEFINITION 6.2.1. A number n in W is said to be **prime** if and only if $n > 1$ and its only factors are n and 1.

DEFINITION 6.2.2. A number n in W is said to be **composite** if and only if $n > 1$ and n is not prime.

From the above definitions it is easy to determine that the first five prime numbers in W are 2, 3, 5, 7, and 11, while the first five composite numbers in W are 4, 6, 8, 9, and 10. Noticeably absent from the sets of prime and composite numbers are the whole numbers 0 and 1. The numbers 0 and 1 are neither prime nor composite. An alternative way of defining a prime number is to say that it is a whole number with exactly two *distinct* factors. Zero, of course, has every whole number as a factor; $0 = 0 \cdot 0 = 0 \cdot 1 = 0 \cdot 2 = 0 \cdot 3$, and so on. The number 1 has two factors since $1 = 1 \cdot 1$, but only one *distinct* factor since $1 = 1$.

Many students wonder why prime numbers are defined in such a way as to exclude the number 1. If we were to remove the restriction $n > 1$ from Definition 6.2.1, the number 1 would qualify as a prime since its only factors are itself (1) and 1. One reason for excluding 1 as a prime is that many theorems can be proved which hold for the primes but not for the composite numbers or for 0 and 1; the Fundamental Theorem of Arithmetic discussed in the next section is an example. If 1 were included as a prime, it would be necessary to say that such theorems are true for all primes p, $p \neq 1$.

Let us mention here that we could define prime and composite numbers to include the negative integers by saying that a negative integer is prime if its additive inverse is prime, and that a negative integer is composite if it is not a prime.

It would be very convenient to have a mathematical formula that could be used to generate in succession all the primes. However, no one has yet produced such a formula although many have tried. Some expressions, such as $n^2 - n + 41$, may be used to generate a large number of primes. The expression $n^2 - n + 41$ will yield a prime number for every positive integer n such that $1 \leq n \leq 40$. For $n = 1, 2, \ldots , 40$, we get the numbers 41, 43, 47, 53, 61, 71, 83, and so on—all of which are prime. However, if $n = 41$, the result is $41^2 - 41 + 41 = 41^2 = 41 \cdot 41$, and $41 \cdot 41$ is obviously not prime. The student who devises a formula for generating all the primes will live in history. However, we warn the prospective inventor that most mathematicians believe no such formula is possible.

Fermat suggested that while the expression $2^{2^n} + 1$ would not generate *all* of the primes, it would generate *only* primes for positive integral values of n. Although Fermat's conjecture was wrong, numbers of this form are nevertheless called **Fermat numbers**. If $n = 0, 1, 2, 3$, and 4, we obtain respectively the primes 3, 5, 17, 257, and 65,537. However, Leonard Euler (1707–1783), a Swiss mathematician, showed that for $n = 5$ the resulting Fermat number is $2^{2^5} + 1 = 2^{32} + 1 = 4,294,967,297$; and since $4,294,967,297 = 641 \times 6,700,417$, it is not prime. Whether or not $2^{2^n} + 1$ is prime for any value of $n > 4$ is not known; however, well over 40 composite Fermat numbers are known and it is now conjectured that all Fermat numbers for $n > 4$ are composite.

One rather crude but effective method for determining all the primes less than or equal to a given number n is to use the "Sieve of Eratosthenes." Eratosthenes (about 230 B.C.) was a Greek mathematician. The reason for calling his method of finding primes a sieve will be obvious as soon as the procedure is explained. Suppose, for example, we wish to find all the primes less than or equal to 50. Write down the numbers from 1 to 50 and immediately strike out 1, which is not a prime. Then circle 2 (the first prime) and strike out all multiples of 2 greater than 2: 4, 6, 8, and so on. We then have:

$$
\begin{array}{cccccccccc}
\cancel{1} & \boxed{2} & 3 & \cancel{4} & 5 & \cancel{6} & 7 & \cancel{8} & 9 & \cancel{10} \\
11 & \cancel{12} & 13 & \cancel{14} & 15 & \cancel{16} & 17 & \cancel{18} & 19 & \cancel{20} \\
21 & \cancel{22} & 23 & \cancel{24} & 25 & \cancel{26} & 27 & \cancel{28} & 29 & \cancel{30} \\
31 & \cancel{32} & 33 & \cancel{34} & 35 & \cancel{36} & 37 & \cancel{38} & 39 & \cancel{40} \\
41 & \cancel{42} & 43 & \cancel{44} & 45 & \cancel{46} & 47 & \cancel{48} & 49 & \cancel{50}
\end{array}
$$

Next circle 3 and strike out all of its multiples greater than 3. (Some multiples of 3 will have already been eliminated as multiples of 2.) Next circle 5 and continue this process. When finished, all the prime numbers less than or equal to 50 will be circled as follows:

$\cancel{1}$　②　③　$\cancel{4}$　⑤　$\cancel{6}$　⑦　$\cancel{8}$　$\cancel{9}$　$\cancel{10}$

⑪　$\cancel{12}$　⑬　$\cancel{14}$　$\cancel{15}$　$\cancel{16}$　⑰　$\cancel{18}$　⑲　$\cancel{20}$

$\cancel{21}$　$\cancel{22}$　㉓　$\cancel{24}$　$\cancel{25}$　$\cancel{26}$　$\cancel{27}$　$\cancel{28}$　㉙　$\cancel{30}$

㉛　$\cancel{32}$　$\cancel{33}$　$\cancel{34}$　$\cancel{35}$　$\cancel{36}$　㊲　$\cancel{38}$　$\cancel{39}$　$\cancel{40}$

㊶　$\cancel{42}$　㊸　$\cancel{44}$　$\cancel{45}$　$\cancel{46}$　㊼　$\cancel{48}$　$\cancel{49}$　$\cancel{50}$

There are many interesting and easily stated problems concerning prime numbers that have not been solved. We do not know, for example, whether there is a greatest pair of *twin primes*. **Twin primes** are pairs of primes whose difference is 2; a few examples are $(3, 5)$, $(5, 7)$, $(11, 13)$, $(41, 43)$, and $(59, 61)$. Some larger twin primes are $(2081, 2083)$, $(2129, 2131)$, and $(2711, 2713)$. Mathematicians have found that no matter how far a table of prime numbers is extended, twin primes occur every now and again. By inductive reasoning, then, it is believed that there is no greatest pair of twin primes.

A closely related question concerns whether or not there can be an arbitrarily long finite sequence of consecutive composite numbers before the next prime occurs. The answer to this question is yes; for, given any arbitrary natural number n, the n numbers running from $(n + 1)! + 2$ to $(n + 1)! + n + 1$ will all be composite. With a little effort, you can probably justify this statement yourself. (If you do not understand the symbolism $(n + 1)!$, consult the index for *factorial notation*.)

Example 1.   Find a sequence of five consecutive composite numbers.

Solution: The sequence from $(n + 1)! + 2$ to $(n + 1)! + n + 1$ will result in such a sequence if $n = 5$. Hence the first term of the sequence is $(5 + 1)! + 2 = 6! + 2$, and the last term is $(5 + 1)! + 5 + 1 = 6! + 6$. The complete sequence is, therefore: $(6! + 2)$, $(6! + 3)$, $(6! + 4)$, $(6! + 5)$, and $(6! + 6)$. Since $6! = 6 \cdot 5 \cdot 4 \cdot 3 \cdot 2 \cdot 1$, it is obvious from the distributive property that the sum $(6! + 2)$ has a factor of 2. Similarly, $(6! + 3)$ has a factor of 3, $(6! + 4)$ has a factor of 4, and so on. Hence all of the numbers of the sequence are composite. The numbers of the sequence simplify to 722, 723, 724, 725, and 726.

Although this is not the first such sequence of 5 composite numbers, the important point is that an arbitrarily long sequence can always be found using the given formula. As a matter of interest, the first sequence of five composite numbers is: 24, 25, 26, 27, and 28.

In 1742 the otherwise obscure mathematician Christian Goldbach (1690–1764) conjectured that every even integer greater than 4 may be expressed as the sum of two odd primes; for example, $16 = 5 + 11 = 3 + 13$. Goldbach also conjectured that every odd number greater than 7 may be expressed as the sum of three odd primes; an example is $23 = 3 + 7 + 13 = 5 + 7 + 11$. To date, no one has been able to either prove or disprove Goldbach's conjectures. It seems that some of the questions that can be stated most simply are among the most difficult to prove.

Euclid (about 300 B.C.) proved the important fact that no greatest prime exists and hence there are infinitely many primes. His method of proof is slightly different from those previously presented in this text. Euclid reasoned as follows: Suppose there is a finite set of primes $\{p_1, p_2, p_3, p_4, \ldots, p_n\}$, where p_n is the greatest prime. Consider the number $M = p_1 \cdot p_2 \cdot p_3 \cdot p_4 \cdot \cdots \cdot p_n + 1$. Note that M is the sum of two addends; the first addend is the product of all the primes and the second addend is 1. Now M must be either prime or composite. If M is prime, then there is a larger prime than p_n since $M > p_n$. This contradicts the supposition that p_n is the greatest prime. If, however, M is composite, then it must be divisible by some prime (as we shall see in Section 6.3). But M is not divisible by any of the primes $p_1, p_2, p_3, \ldots, p_n$ since there will be a remainder of 1 in each case, and hence M must be divisible by some prime greater than p_n. This also contradicts the supposition that p_n is the greatest prime. In either case, there is no greatest prime and hence it follows that there are infinitely many primes.

A large number known to be prime is $2^{11213} - 1$, which has 3376 digits and was discovered in 1963 by Donald Gillies at the University of Illinois. Currently, the largest known prime is $2^{19937} - 1$, which was reported by Bryant Tuckerman of I.B.M. Corporation in 1971. This number has 6002 digits; it was discovered by using an IBM 360/91 computer. Euclid's proof, however, assures us that even this very large number is not the greatest prime! Finally, we note that if there were a greatest prime, there would necessarily be a greatest pair of twin primes. If there were a greatest pair of twin primes, would there necessarily be a greatest prime?

EXERCISE 6.2

1. Define a *prime* number.

2. Define a *composite* number.

3. Using the Sieve of Eratosthenes:
   (a) Find all primes less than 100.
   (b) Identify all twin primes less than 100.

4. Show that each of the following even numbers may be written as the sum of two (not necessarily distinct) odd primes: 6, 8, 10, 12, 72, 84.

5. Write 22 as the sum of two primes in as many ways as possible.

6. Show that each of the following odd numbers may be written as the sum of three (not necessarily distinct) odd primes: 9, 11, 13, 15, 35.

7. Find the first four Fermat numbers by letting $n = 0, 1, 2$, and 3. (Note that exponentiation is not associative and that $2^{2^n} + 1 = 2^{(2^n)} + 1$ rather than $(2^2)^n + 1$.)

8. (a) By the method of Example 1 of this section, find seven consecutive composite numbers.

   (b) What are the first seven consecutive composite numbers? Each is less than 100. [*Hint:* See Problem 3.]

6.3 The Fundamental Theorem of Arithmetic

Composite numbers may be expressed as the product of factors. For example: $12 = 1 \cdot 12$, $12 = 2 \cdot 6$, $12 = 3 \cdot 4$, and $12 = 2 \cdot 2 \cdot 3$. Each of these indicated products is called a **factorization** of the number 12. Every composite number may be expressed as the product of prime numbers by selecting any factorization of the number that does not include 1 and continuing to factor those factors that are not already prime. Consider 24 as an example: $24 = 3 \cdot 8$; 3 is a prime but 8 is not and can therefore be factored. So we write $24 = 3 \cdot 8 = 3 \cdot 2 \cdot 4$. Now we see that all factors are prime except 4. Continuing, we finally obtain $24 = 3 \cdot 8 = 3 \cdot 2 \cdot 4 = 3 \cdot 2 \cdot 2 \cdot 2$. The last expression, $3 \cdot 2 \cdot 2 \cdot 2$, is called the **prime factorization** or the **complete factorization** of the number 24 since all the factors are prime. It is common practice to arrange the prime factors in order from least to greatest and to write $24 = 2 \cdot 2 \cdot 2 \cdot 3 = 2^3 \cdot 3$.

Another method for factoring a composite number into prime factors is to test the primes as divisors in succession beginning with the least prime: 2, 3, 5, 7, 11, 13, and so on. Continue dividing by each prime as many times as possible before going to the next. Several illustrations follow.

Example 1. Factor 48 into prime factors.

Solution:

$$48 = 2 \cdot 24$$
$$= 2 \cdot 2 \cdot 12$$
$$= 2 \cdot 2 \cdot 2 \cdot 6$$
$$= 2 \cdot 2 \cdot 2 \cdot 2 \cdot 3$$
$$= 2^4 \cdot 3$$

$$\begin{array}{r} 2)\overline{48} \\ 2)\overline{24} \\ 2)\overline{12} \\ 2)\overline{6} \\ 3 \end{array}$$

Example 2. Find the prime factorization of 700.

Solution:

$$700 = 2 \cdot 350$$
$$= 2 \cdot 2 \cdot 175$$
$$= 2 \cdot 2 \cdot 5 \cdot 35$$
$$= 2 \cdot 2 \cdot 5 \cdot 5 \cdot 7$$
$$= 2^2 \cdot 5^2 \cdot 7$$

$$2)\overline{700}$$
$$2)\overline{350}$$
$$5)\overline{175}$$
$$5)\underline{35}$$
$$7$$

Example 3. Find the complete factorization of 1547.

Solution:

$$1547 = 7 \cdot 221$$
$$= 7 \cdot 13 \cdot 17$$

$$7)\overline{1547}$$
$$13)\underline{221}$$
$$17$$

Factoring very large numbers into primes can become tedious and is best accomplished by using modern high-speed computers. Finding the prime factors of 2,839,003 (743 and 3821) might prove to be less than exciting. Of course, if we test for prime divisors of 743 and 3821, there will be none except the numbers themselves since 743 and 3821 are prime.

In searching for the prime factors of a given number, it is useful to know that we need try no primes larger than the square root of the number. To see how this works, let us examine, for example, all the pairs of factors of 100.

$$1 \times 100$$
$$2 \times 50$$
$$4 \times 25$$
$$5 \times 20$$
$$10 \times 10 \quad \longleftarrow$$
$$20 \times 5$$
$$25 \times 4$$
$$50 \times 2$$
$$100 \times 1$$

As larger numbers are used for the first factor of 100, smaller numbers must be used for the second factor. The two factors are equal when we reach 10 (the square root of 100). From the list of factors it is apparent that all pairs of factors following 10×10 may be found simply by commuting previous pairs of factors. Hence there is no need to test numbers larger than 10 as factors of 100 since they were already found by testing smaller numbers.

Example 4. Find the prime factorization of 143.

Solution: Since $11^2 = 121$ and $12^2 = 144$, we see that $11 < \sqrt{143} < 12$. Therefore, we need to test as divisors only prime numbers less than 12. In succession we test 2, 3, 5, 7, and 11. Upon dividing by 11 we find that $143 = 11 \cdot 13$; and since 13 is also prime, the factorization is complete.

Example 5. Factor 819 into prime factors.

Solution: Since $28^2 = 784$ and $29^2 = 841$, primes to be tested are those less than 29: 2, 3, 5, 7, 11, 13, 17, 19, and 23. Upon dividing by 3 we find that $819 = 3 \cdot 273$. Now, since $16 < \sqrt{273} < 17$, it is necessary to consider only those primes less than 17: 2, 3, 5, 7, 11, and 13. However, 3 is again a divisor and $819 = 3 \cdot 3 \cdot 91$. Now it is necessary to test only primes less than 10, and the result is $819 = 3 \cdot 3 \cdot 7 \cdot 13$. All these factors are prime and the factorization is complete.

Example 6. Find the complete factorization of 743.

Solution: Since $27^2 = 729$ and $28^2 = 784$, only primes less than 28 need to be considered: 2, 3, 5, 7, 11, 13, 17, 19, and 23. Since none of these divides 743, we conclude that 743 is prime.

The following theorem summarizes the discussion of this section. Although the title "Unique Prime Factorization Theorem" best describes the theorem, it is usually referred to as "The Fundamental Theorem of Arithmetic" because of its great importance.

THEOREM 6.3.1. **The Fundamental Theorem of Arithmetic** (Unique Prime Factorization Theorem). Every composite number can be expressed uniquely as the product of prime factors, if the order of the factors is disregarded.

The proof of the Fundamental Theorem of Arithmetic may be found in almost any text on the theory of numbers.

EXERCISE 6.3

1. Find the prime factorization of the following numbers:
   (a) 60   (b) 108   (c) 51
   (d) 156   (e) 510   (f) 362

2. Attempt to find two different prime factorizations of 24 by starting first with $24 = 8 \times 3$ and then with $24 = 6 \times 4$.

3. Factor -18 uniquely as the product of -1 and prime factors.

4. A *perfect number P* is a number that equals the sum of its distinct factors, excluding P itself. For example, 6 is a perfect number since its factors

are $\{1, 2, 3, 6\}$ and $6 = 1 + 2 + 3$. Two other perfect numbers are 496 and 8128. There is only one two-digit perfect number. Try to find it. [*Hint:* It is less than 40.]

5. A number is said to be a *cubic number* if it is equal to x^3 where $x \in N$. The first four cubic numbers are $1 = 1^3$, $8 = 2^3$, $27 = 3^3$, and $64 = 4^3$. The smallest number that may be written as the sum of two cubic numbers in two different ways is 1729. It is the sum of 12^3 and 1^3. See if you can find the other two cubic numbers whose sum is 1729.

6.4 Odd and Even Numbers

Odd and even numbers have already been referred to in our discussions since we assumed that most students have some knowledge of them. However, we now give formal definitions of odd and even numbers.

DEFINITION 6.4.1. A number n in W is said to be an **even number** if and only if it has a factor of 2.

The even numbers, then, are the members of $E = \{0, 2, 4, 6, 8, \ldots\}$. A similar definition can be given for the integers. The even integers are the members of $\{\ldots, -6, -4, -2, 0, 2, 4, 6, \ldots\}$.

DEFINITION 6.4.2. A number n in W is said to be an **odd number** if it is not an even number.

The definition of an odd number implies that it is a member of the set $F = \{1, 3, 5, 7, \ldots\}$. Extending the definition to the integers, the odd integers are the members of $\{\ldots, -5, -3, -1, 1, 3, 5, \ldots\}$.

The even whole numbers may be identified as the set $E = \{x \mid x = 2n, n \in W\}$. Similarly, the set of odd whole numbers may be indicated by $F = \{x \mid x = 2n + 1, n \in W\}$. As will be seen in the examples below, it is frequently convenient to represent an even number in the form $2n$ and an odd number in the form $2n + 1$. Unless otherwise stated, we shall use the terms odd and even numbers to refer to the odd and even whole numbers rather than to the integers.

Example 1. Prove the closure property holds for addition in the set of even numbers.

Solution: Let $2a$ and $2b$ represent any two even numbers. Then their sum $2a + 2b = 2(a + b)$ by the distributive property. Since a and b are in W and since the closure property holds for addition in W, we can write $a + b = c$ where $c \in W$. Then $2(a + b) = 2c$, but $2c$ is by definition an even number. Therefore, the sum of any two even numbers is an even number and the closure property holds for addition in the even numbers.

Example 2. Prove the closure property does *not* hold for addition in the set of odd numbers.

Solution: One counterexample is sufficient to demonstrate that the closure property does not hold for addition in the odd numbers. Since $7 = 2(3) + 1$ and $9 = 2(4) + 1$, 7 and 9 are by definition odd numbers. However, their sum $7 + 9 = 16 = 2(8)$, which we recognize as an even number. Therefore, the set of odd numbers is not closed under addition.

EXERCISE 6.4

1. Prove that the sum of any two odd numbers is an even number. [*Hint:* Let $(2a + 1)$ and $(2b + 1)$ represent any two odd numbers.]

2. Prove that the sum of any even number and any odd number is an odd number.

3. Prove that the set of odd numbers is closed with respect to the operation of multiplication.

4. Prove that the product of any even number and any odd number is an even number.

5. If a and b are twin primes, then $ab + 1$ is a perfect square.
   (a) Verify this statement by checking the first three pairs of twin primes.
   (b) Prove the given statement. [*Hint:* Let $(x - 1)$ and $(x + 1)$ represent any pair of twin primes.]

6. Consider the following:

$$1 = 1^2$$
$$1 + 3 = 4 = 2^2$$
$$1 + 3 + 5 = 9 = 3^2$$
$$1 + 3 + 5 + 7 = 16 = 4^2$$

   (a) By inductive reasoning, what would you expect to be the sum of $1 + 3 + 5 + 7 + 9$?
   (b) Can you suggest a general mathematical expression for the sum of the first n odd numbers? Test your expression by using a few examples.

7. Why must a number be a perfect square if it has an odd number of distinct factors?

6.5 Special Numbers with Interesting Properties

In Problem 4 of Exercise 6.3, we defined a **perfect number** P as a number that is equal to the sum of its proper divisors. The proper divisors of a number exclude the number itself. The smallest perfect number is 6, and we see that $6 = 1 + 2 + 3$. The next perfect number is 28, and again it is a simple matter to verify that 28 is equal to the sum of its proper divisors:

$28 = 1 + 2 + 4 + 7 + 14$. The first six perfect numbers are: 6; 28; 496; 8128; 33,550,336; and 8,589,869,056. A book which the author of this text studied from as a college student stated that 130,816 and 2,096,128 are perfect numbers. However, this has been proved to be false. One might expect to find perfect numbers between 8128 and 33,550,336 since the gap is so large, but there are none.

Perfect numbers have intrigued man for hundreds of years. The famous Greek mathematician Euclid (about 300 B.C.) was certainly familiar with the first three or four perfect numbers. People observed that the moon completes its path around the earth in 28 days and that 28 is a perfect number. Aurelius Augustus (354–430) remarked that God effected the creation in 6 days because 6 is a perfect number. This line of reasoning was extended by Alcuin (735–804) of York and Tours, who explained that since there were 8 souls in Noah's ark from which sprang the entire human race, the second creation was less perfect than the first. His conclusion rested on the fact that 8 is not a perfect number, but rather it is deficient. The number 8 is said to be deficient because the sum of its proper divisors $1 + 2 + 4 = 7$, which is less than 8. Similarly, we say that the number 12 is abundant since the sum of its proper divisors $1 + 2 + 3 + 4 + 6 = 16$, which is greater than 12.

How many perfect numbers are there? No one knows; but we suspect there are infinitely many. Are all perfect numbers even? Again, we do not know; but we suspect the answer is yes since no one has been able to find an odd perfect number.

There are 24 known perfect numbers. All of these are of the form $2^{p-1}(2^p - 1)$, where both p and $(2^p - 1)$ are prime. The known values for p that yield perfect numbers are

2, 3, 5, 7, 13, 17, 19, 31, 61, 89, 107, 127, 521, 607, 1279, 2203, 2281, 3217, 4253, 4423, 9689, 9941, 11213, and 19937.

The list, of course, is growing with the aid of the electronic computer. In Section 6.2 we mentioned that $2^{19937} - 1$ is the largest known prime. It follows immediately that $(2^{19936})(2^{19937} - 1)$, which has 12,003 digits, is the largest known perfect number.

Although much about perfect numbers remains unknown, the following theorem has been proved:

THEOREM 6.5.1.

  (i) The number $2^{p-1}(2^p - 1)$ is perfect if and only if p is prime and $2^p - 1$ is prime.

 (ii) An even number is perfect if and only if it can be expressed as $2^{p-1}(2^p - 1)$ where p is prime and $2^p - 1$ is prime.

Mathematicians are quite certain that if any perfect numbers are discovered in the future, they will be very large numbers. Since it has been demonstrated that there is no greatest prime, we suspect there is no greatest perfect number; but this has not been proved.

Example 1. Let $p = 7$ in the expression $2^{p-1}(2^p - 1)$ and find the value of the fourth perfect number.

Solution: Note that $2^7 - 1 = 63$, which is prime. Therefore, Theorem 6.5.1(i) applies and we see that:

$$
\begin{aligned}
2^{p-1}(2^p - 1) &= 2^{7-1}(2^7 - 1) \\
&= 2^6(2^7 - 1) \\
&= 64(128 - 1) \\
&= 64(127) \\
&= 8128
\end{aligned}
$$

Next, let us consider the pair of numbers 220 and 284. The sum of the proper divisors of 220 is 284, while the sum of the proper divisors of 284 is 220. Such pairs of numbers are called amicable (meaning friendly). The smallest pair of amicable numbers is 220 and 284. Other examples are (1184; 1210), (17,296; 18,416) and (9,363,584; 9,437,056).

As in the case of perfect numbers, men of long ago knew that amicable numbers existed. Some people even believed that using amicable numbers might help establish a close friendship with another person. In Genesis XXXII:14, we read that among Jacob's presents to his brother Esau were 220 goats (200 she goats and 20 he goats) as well as 220 sheep (200 ewes and 20 rams). Interestingly enough, Jacob did recapture the good will of his brother Esau following the presentation of such gifts.

In 1634 the French mathematician Mersenne (1588–1648) remarked that "220 and 284 can signify the perfect friendship." In 1636 Mersenne correctly stated that Fermat had found a second pair of amicable numbers 17,296 and 18,416. He also stated that Fermat had discovered a general technique for identifying certain pairs of amicable numbers. Others, however, give Descartes credit for the technique. In any event, over 900 pairs of amicable numbers have been identified, all of which are quite large. Euler alone identified about 60 pairs of amicable numbers. The relatively small pair, 1184 and 1210, which had been missed by professional mathematicians, was discovered by a 16-year-old Italian boy, Nicola Paganini, in 1866. There are evidently infinitely many pairs of amicable numbers, although this has not yet been proved. Despite special techniques that have been developed for identifying certain pairs of amicable numbers, no general expression exists for identifying *all* pairs of amicable numbers.

We should not conclude this section of the text without mentioning Fermat's most famous theorem. The equation $x^2 + y^2 = z^2$ (which is, of course, an expression of the Pythagorean Theorem) has infinitely many positive integral solutions, such as $3^2 + 4^2 = 5^2$ and $5^2 + 12^2 = 13^2$. Sets of integers for which this equation is true are called *Pythagorean triples*. Hence 3, 4, and 5 as well as 5, 12, and 13 are Pythagorean triples. It is natural to wonder whether or not there are any triples that will make the equation $x^n + y^n = z^n$ true if $n > 2$. There are no known positive integral solutions for $n > 2$, and Fermat claimed in 1637 to have proved that, in fact, none exist. However, he did not show his proof and his statement has become known as "Fermat's Last Theorem." In his copy of Diophantus' *Arithmetica* Fermat wrote, "I have discovered a truly remarkable proof but the margin is too small to contain it." To this day, except for special cases, the problem remains unsolved. For this reason, perhaps, we should call Fermat's Last Theorem a conjecture rather than a theorem.

There are many unsolved classical problems in number theory; there are also many new problems that are being formulated and solved by today's mathematicians. Although the number of direct practical applications of number theory has been somewhat limited, the stimulus it has provided in the study of other branches of mathematics has been quite rewarding. Some knowledge of number theory is, of course, essential in the design and construction of electronic computers.

 Diophantine Equations

One of the oldest and most interesting topics in number theory is that of **Diophantine equations**. Although Diophantus lived about 250 A.D. and probably wrote 12 or 13 books, it was not until about 1200 years later that his work began to attract the attention of prominent European mathematicians. Only six of Diophantus' books on *arithmetica* have survived along with a few other fragmentary writings. *Arithmetica* was the Greek word for the *theory of numbers*. (What we call *arithmetic* was called *logistica* by the Greeks.)

Diophantine problems usually require integral solutions of equations, and frequently multiple solutions are possible. For example, suppose we ask a cashier to exchange a quarter for nickels and dimes. How can this be done? Obviously, the quarter could be exchanged for any of the following:

> 1 nickel and 2 dimes
> 3 nickels and 1 dime
> 5 nickels and 0 dimes

If we try to solve this problem with an equation, and let n = the number of nickels and d = the number of dimes, we have $5n + 10d = 25$. This Diophantine equation has many solutions but only three nonnegative integral solutions. If we place another restriction on the original problem by requesting as few coins as possible in exchange for the quarter, then there is a unique solution: 1 nickel and 2 dimes.

Although general solutions have been formulated for certain types of Diophantine equations, we will not discuss them here. Instead, we shall describe the well-known and somewhat intriguing problem of the three shipwrecked sailors; the student may have the fun of attempting to solve it without assistance.

Three sailors were cast upon an island where they could find no food other than coconuts. They gathered all the coconuts they could find and put them in a pile. They were to divide the coconuts into three equal shares the following day. However, during the night, one sailor awakened and decided to take his share of the coconuts and hide them. Upon attempting to divide them into three equal groups, one coconut remained; he threw it into the sea and buried his third of those that were left. Later, another of the three sailors awakened and discovered that by throwing away one coconut, those remaining were divisible by 3. He likewise buried a third of the leftover coconuts. Finally, the third sailor awakened and went through exactly the same procedure. If there was a whole number of coconuts left in the morning, what is the *smallest* number of coconuts the sailors could have collected?

Although we have mentioned just a few simple problems to introduce Diophantine equations, do not be misled into believing that it is a trivial topic. Many important concepts are embraced in the study of Diophantine equations.

In general, Diophantine problems result in more unknowns than equations. The problem involving change for a quarter resulted in $5n + 10d = 25$, which has two unknowns in only one equation. Solutions to the problem are restricted, however, since n and d must be integers. In more complex problems, we might be involved with many equations, but continue to have the number of unknowns exceed the number of equations. In fact, Fermat's Last Theorem is a Diophantine problem concerned with one equation having three unknowns.

EXERCISE 6.6

1. Define a *perfect number*.

2. Find the first four perfect numbers by letting $p = 2, 3, 5,$ and 7 in the expression $2^{p-1}(2^p - 1)$.

3. (a) How many perfect numbers are there?
   (b) Are there any odd perfect numbers?

4. (a) Do you believe perfect numbers have any religious or mystical significance?
   (b) Is there any such thing as a "lucky" or "unlucky" number?
   (c) If you were building a hotel, would you have a thirteenth floor? Explain.
   (d) What significance do you attach to the fact that the manned space vehicle named Apollo 13 did not complete its mission?

5. How are the numbers a and b related if they are an *amicable* pair?

6. (a) What are the two smallest amicable pairs of numbers?
   (b) Are there infinitely many pairs of amicable numbers?

7. (a) The expression $n^2 + (\frac{n^2-1}{2})^2 = (\frac{n^2+1}{2})^2$ will determine a set of Pythagorean triples for each *odd* integer $n > 1$. (For example, if $n = 3$, we get $3^2 + 4^2 = 5^2$, and $\{3, 4, 5\}$ is a Pythagorean triple.) Find the first four sets of Pythagorean triples that may be generated with this equation by letting $n = 3, 5, 7$, and 9.
   (b) The expression $n^2 + (\frac{n^2-4}{4})^2 = (\frac{n^2+4}{4})^2$ will determine a set of Pythagorean triples for each *even* integer $n > 2$. Find the first four sets of Pythagorean triples that may be generated with this equation by letting $n = 4, 6, 8$, and 10.

8. Solve the problem concerning the shipwrecked sailors mentioned in Section 6.6.

9. A problem of "ancient vintage" concerns a wine merchant who has an 8-gallon jug of wine. A customer wishes to buy 4 gallons, but only has an empty 3-gallon jug and an empty 5-gallon jug. How can the merchant sell his customer exactly 4 gallons by using only the jugs that have been mentioned?

REVIEW EXERCISES

1. Name the five smallest prime numbers.

2. Name the five smallest composite numbers.

3. With reference to being either prime or composite, what is the status of 0 and 1? Explain.

4. Find the prime factorization of:
   (a) 1864   (b) 7240

5. If 131 has no prime factors less than 12, then 131 itself must be prime. Why is this true? Is 131 prime? Explain.

6. (a) What is the first sequence of three consecutive composite numbers?
   (b) What is the first sequence of five consecutive composite numbers?

7. What is the smallest abundant number?

8. What is the smallest abundant number that is greater than 12?

9. Write 24 as the sum of two (not necessarily distinct) odd primes in as many ways as possible.

10. Write 29 as the sum of 3 (not necessarily distinct) odd primes in as many ways as possible.

Roberto Beredecio, *The Cube and the Perspective*
1935. Duco airbrushed on steel panel mounted on wood. 30 x 26″.
Collection, The Museum of Modern Art, New York

FINITE MATHEMATICAL SYSTEMS AND THE CONGRUENCE RELATION

7.1 Introduction

The various mathematical systems that have been discussed, as well as those to be considered in this chapter, may be classified according to their common properties. Although we can fabricate many mathematical systems, the most familiar of these have many properties in common and the well-known classifications of *group*, *ring*, and *field* will embrace the majority of them.

Initially, the person with a "practical" orientation is likely to believe that almost no applications could be found for these abstract concepts of groups, rings, and fields. In fact, many mathematicians and other scientists were, at first, of this opinion. Today, however, we find that many applications have resulted from the study of various mathematical structures. The theory of groups, for example, is used in studying the physical structure and nature of matter; Albert Einstein found group theory necessary for his special theory of relativity. Also, group theory is used in studying the nature of human genes, the elements by which hereditary characteristics are determined and transmitted. Over and over again, history has demonstrated that we can never be certain where a new mathematical theory will lead us or what applications may result from it.

159

The earliest work on group theory was done by the Frenchman Augustin-Louis Cauchy (1789–1857). The great Norwegian mathematician Niels Henrik Abel (1802–1829) added to the knowledge of group theory and used it to prove that there is no general solution to a fifth-degree polynomial equation. This problem had bothered mathematicians for many years since solutions had been found for equations of degrees 1, 2, 3, and 4. It seemed (falsely) that only a matter of time and effort would be necessary to find similar solutions for higher-degree equations; many hours were spent in this attempt. Unfortunately, Abel's contributions to mathematics ended abruptly with his early death from tuberculosis. Abel had been poor all his life and it was ironic that notice of his appointment as a professor at the University of Berlin should arrive in the mail just after his death.

No discussion of group theory would be complete without mention of the French mathematician Evariste Galois (1811–1832). All of his short life, Galois was plagued by "bucking the management." He had difficulty getting along with his teachers and with governmental officials. He considered others intellectually inferior to himself; indeed, most of them probably were since Galois was, no doubt, a genius. However, it is unfortunate that he did not use his intelligence to cope with life in a more realistic fashion and at least stay alive. Prior to his twenty-first birthday, Galois agreed to fight a duel as a matter of "honor." The matter concerned his one and only love affair with a girl named Eve. There is reason to believe that the entire affair, including his relationship with Eve, was staged by politicians who wanted to eliminate Galois as a revolutionary. Galois knew he would be killed since he was no match for his opponent. Fortunately, Galois spent most of the night before the duel writing his mathematical theories on groups and gathering together other mathematical notes. The total of his works amounts to about 60 printed pages. Had Galois lived a normal life span, he would probably have been one of the most productive mathematicians of all time. One might wish that Galois had stayed with his mathematics and let the politicians fight their own duels.

In studying the properties of groups, rings, and fields, we will find it convenient to use certain finite mathematical systems, which are discussed in the next section. Although the numerous applications of the introductory concepts developed in this chapter are not self-evident, it is comforting to the practical man to know that they exist.

7.2 Finite Mathematical Systems

In Chapter 5 we identified and discussed several number systems including the system of whole numbers, the system of integers, and the system of rational numbers. Each of these systems is concerned with a set of infinitely many elements. Mathematical systems may also be constructed using a finite number of elements; such systems are appropriately called finite systems.

It is rather easy to create a finite mathematical system with only a few elements. We shall construct a mathematical system with four elements by referring to a "four-hour clock" as a model; see Figure 7.2.1. Once the clock arithmetic system has been established, however, we will no longer need to refer to the model from which it was abstracted.

From the figure we see that in one hour the hand on the clock will move from 0 to 1, in two hours from 0 to 2, and so on. In four hours the hand will have made one complete revolution and will be back to 0 again. Note that the origin, or starting point, on the clock could have been named either 0 or 4. On a conventional twelve-hour clock, the origin is designated as 12; it could equally well be called 0.

We have identified the set of elements in our finite system as $\{0, 1, 2, 3\}$. Next, we shall define a binary operation on the set—addition $(+)$. Suppose we wish to find the sum $1 + 2$. We place the hand on 1 and move it in a clockwise direction 2 units; it will then be at 3 and we see that $1 + 2 = 3$. Similarly, if we wish to find the sum $2 + 3$, we place the hand on 2 and move it 3 units in a clockwise direction and find that $2 + 3 = 1$. In general, then, if we wish to find the sum $a + b$, we place the hand on a and move it clockwise b units. The reading taken at this position is the sum $a + b$. The fact that $2 + 3 = 1$ in a four-hour clock arithmetic should not seem particularly strange to those who are accustomed to twelve-hour clocks. If it is 11 o'clock, using a twelve-hour clock, four hours later it will be 3 o'clock. In this case $11 + 4 = 3$.

In constructing an addition table for a four-hour clock arithmetic, it is necessary to consider only the elements 0, 1, 2, and 3. Using the method for finding sums described above, we can construct an addition table as shown in Figure 7.2.2. Check the results and see if you agree that they are correct.

We now have a set of elements for our mathematical system and an operation, addition ($+$), defined on the set. What are some of the properties of this mathematical system?

By examining the addition table in Figure 7.2.2, we see that the closure property for addition is satisfied since for each ordered pair of elements (a, b) in the set, there exists a unique element $a + b$ in the set.

FIGURE 7.2.2

+	0	1	2	3
0	0	1	2	3
1	1	2	3	0
2	2	3	0	1
3	3	0	1	2

The commutative property also holds since for all (a, b) in the set, $a + b = b + a$. We can show this without testing all possible sums in the table. The table is symmetric about the diagonal extending from the upper left corner to the lower right corner; and since each sum $a + b$ is reflected across this diagonal to $b + a$, symmetry indicates that $a + b = b + a$.

It turns out that the associative property holds for addition, although quite a few tests must be made to exhaust all possibilities. The truth of the associative property may be verified (but not proved) by testing several cases at random to see if $(a + b) + c = a + (b + c)$.

An identity element for addition exists and is the number 0 (zero). For every element a contained in the set, $a + 0 = 0 + a = a$.

Each element in the set has an additive inverse since for each element a there exists an element $-a$ such that $a + (-a) = 0$. By examining the addition table, it is easy to see that every element has an additive inverse since the identity element 0 appears in every row and column. The inverse of 0 is 0 since $0 + 0 = 0$. The inverse of 1 is 3 since $1 + 3 = 0$. Similarly, the additive inverse of 2 is 2, and the inverse of 3 is 1. It is interesting to note that both 0 and 2 are their own additive inverses, while 1 and 3 are inverses of each other.

As we mentioned in the introduction to the chapter, mathematical systems having certain specific properties are sufficiently important and occur frequently enough to be given special names such as *group*, *ring*, and *field*. A mathematical system such as the one we abstracted by using a four-hour clock is called a *commutative group* (or an *Abelian group* in honor of the Norwegian mathematician Niels Henrik Abel). Without the commutative property, the mathematical system would simply be called a *group*. The properties of groups and some other types of mathematical systems are discussed in Section 7.3.

EXERCISE 7.2

1. (a) Construct an addition table for a three-hour clock arithmetic using elements of {0, 1, 2}.
   (b) Is addition commutative?
   (c) Check several cases such as $(2 + 1) + 2$ and $2 + (1 + 2)$ and indicate whether or not addition seems to be associative.

2. (a) Construct an addition table for a finite mathematical system using a five-hour clock as a model.
   (b) What is the additive inverse of each element of the system?
   (c) What is the identity element for addition?
   (d) Designate the origin on the clock by 5 instead of 0. What is the identity element for addition now?
   (e) What is the identity element for addition using a conventional twelve-hour clock?

3. Why is 3 p.m. sometimes expressed as 1500 hours (particularly by members of the military)? What would 0300 hours mean? 1620 hours?

4. Using a conventional twelve-hour clock, what time will it be in 64 hours if it is now 5 p.m.?

5. Using a five-hour clock, what time will it be in 24 hours if it is now 2 o'clock?

7.3 Commutative Groups, Rings, and Fields

We identified a commutative (Abelian) group in the previous section by using a four-hour clock as a model; now let us give a formal definition of a commutative group.

DEFINITION 7.3.1. A **commutative group** is a mathematical system consisting of a set of elements; one binary operation, addition $(+)$, defined on the set; and the following properties for all a, b, and c in the set:

1. *Closure.* $a + b$ is a unique element of the set.
2. *Commutativity.* $a + b = b + a$.

3. *Associativity.* $(a + b) + c = a + (b + c)$.
4. *Identity Element.* There exists an element i in the set such that $a + i = i + a = a$.
5. *Inverse Elements.* For every a there exists an element $-a$ such that $a + (-a) = i$.

If a mathematical system has all of the properties mentioned above except the commutative property, the system still qualifies as a group but not as a commutative group. Thus we see that commutative groups are a proper subset of all groups.

A mathematical system exhibiting the minimum properties necessary to qualify as a commutative group may have other properties as well. The system we abstracted by using a four-hour clock in Section 7.2 may be used as an example if we also define a second operation. In the four-hour clock arithmetic, a second operation, multiplication ($\cdot$), may be defined on $\{0, 1, 2, 3\}$ quite easily by using the repeated addend approach to multiplication. With this approach and with reference to the addition table in Figure 7.2.2, we find, for example, that $3 \cdot 2 = 2 + 2 + 2 = (2 + 2) + 2 = 0 + 2 = 2$. Using this method to compute all possible products, we can construct a multiplication table as shown in Figure 7.3.1.

FIGURE 7.3.1

$\cdot$	0	1	2	3
0	0	0	0	0
1	0	1	2	3
2	0	2	0	2
3	0	3	2	1

Examining the multiplication table, we see that closure holds since there exists a unique product for every ordered pair (a, b). Since the table is symmetric about a diagonal from the upper left corner to the lower right corner, multiplication is commutative and $a \cdot b = b \cdot a$. The associative property also holds and $(a \cdot b) \cdot c = a \cdot (b \cdot c)$; the reader may wish to check this by testing several cases. Finally, by inspecting the table, we observe that the number 1 is the multiplicative identity for the set.

If a multiplicative inverse existed for each element of the set, then the identity element 1 would appear in every row and column of the multiplication table. In this event, the four-hour clock system would be a commutative group not only under addition but also under the operation of multiplication. However, in examining the multiplication table, we see that neither 0 nor 2 has a multiplicative inverse. The numbers 1 and 3 are multiplicative inverses of themselves since $1 \cdot 1 = 1$ and $3 \cdot 3 = 1$.

Even though the four-hour clock arithmetic does not qualify as a commutative group under multiplication, it more than satisfies the qualifications for being a *ring*. A ring is defined in Definition 7.3.2.

DEFINITION 7.3.2.   A **ring** is a mathematical system consisting of a set of elements; two binary operations, addition $(+)$ and multiplication $(\cdot)$, defined on the set; and the following properties for all a, b, and c in the set:

1. The set is a commutative group under the operation $+$ and therefore has all the properties of a commutative group (Definition 7.3.1).
2. The second operation $\cdot$ has the properties of:
   Closure. $a \cdot b$ is a unique element of the set.
   Associativity. $(a \cdot b) \cdot c = a \cdot (b \cdot c)$.
3. Multiplication $(\cdot)$ is distributive over addition $(+)$.
   Distributive Properties. $a \cdot (b + c) = (a \cdot b) + (a \cdot c)$ and $(b + c) \cdot a = (b \cdot a) + (c \cdot a)$.

In the definition of a ring, both the left and right distributive properties must be mentioned since the operation $\cdot$ is not necessarily commutative. It is interesting to note that the system of integers is a ring. Although the system of integers is a commutative group under addition, it is not a group under multiplication, since not every element has a multiplicative inverse.

Referring again to Section 7.2, we point out that it was not by chance that a four-hour clock was used as a model for a mathematical system; we selected the number 4 because it is a composite number. If a prime number such as 5 had been used in our model, each element except 0 would have had a multiplicative inverse. The addition and multiplication tables for a five-hour clock mathematical system are presented in Figure 7.3.2.

We see from the multiplication table in Figure 7.3.2 that every element except 0 (the additive identity) has a multiplicative inverse. Note that the number 1 appears in every row and column except those headed by 0.

The five-hour clock arithmetic has properties that qualify it as a commutative group under addition; and if the additive identity element 0 is removed from the set, it is also a commutative group under multi-

FIGURE 7.3.2

+	0	1	2	3	4
0	0	1	2	3	4
1	1	2	3	4	0
2	2	3	4	0	1
3	3	4	0	1	2
4	4	0	1	2	3

·	0	1	2	3	4
0	0	0	0	0	0
1	0	1	2	3	4
2	0	2	4	1	3
3	0	3	1	4	2
4	0	4	3	2	1

plication. Furthermore, multiplication is distributive over addition. A system with these properties is called a *field*.

DEFINITION 7.3.3.   A **field** is a mathematical system consisting of a set containing at least two distinct elements; two binary operations, addition ($+$) and multiplication ($\cdot$), defined on the set; and the following properties:

1. The set of elements is a commutative group under addition.
2. The set of elements, with the identity element for addition (0) removed, is a commutative group under multiplication.
3. Multiplication is distributive over additon.

Any finite clock system with a *prime* number of elements is a *field*. The system of rational numbers with infinitely many elements is also a field.

EXERCISE 7.3

1. Define the term *group* as it is used to identify a mathematical system.
2. What is the difference between a *group* and a *commutative group*?
3. What is a *ring*?
4. What is a *field*?
5. Does a four-hour clock number system constitute:
   (a) A group under addition? under multiplication?
   (b) A commutative group under addition? under multiplication?
   (c) A ring under addition and multiplication?
   (d) A field under addition and multiplication?
6. Does a five-hour clock number system constitute:
   (a) A group under addition? under multiplication?
   (b) A commutative group under addition? under multiplication?

(c) A ring under addition and multiplication?

(d) A field under addition and multiplication?

7. (a) What modification in a five-hour clock arithmetic will enable it to then qualify as a commutative group under multiplication?

(b) Is a similar restriction necessary for addition? Explain.

8. (a) How would you define subtraction in a five-hour clock arithmetic?

(b) What would $1 - 4$ equal?

(c) What would $2 - 3$ equal?

(d) What would $0 - 2$ equal?

9. Is the system of integers a ring? a field? Explain.

10. Is the system of rational numbers a ring? a field?

7.4 The Congruence Relation

With reference to a clock arithmetic based on a five-hour clock, it is easy to answer the question, "If it is 4 o'clock, what time will it be 3 hours later?" The answer is $4 + 3 = 2$, as shown in Figure 7.3.2. However, suppose we ask, "If it is 4 o'clock, what time will it be 17 hours from now?" Since the number 17 is not "on the clock," we cannot add 4 and 17 within the system. However, we know that in 5 hours the hand on the clock will be back where it started. Since $17 = 5 + 5 + 5 + 2 = 3(5) + 2$, the hand on the clock will make three complete revolutions and then move two units farther. Therefore, if it is 4 o'clock, 17 hours later the hand will point to $4 + 2 = 1$. Figure 7.4.1 illustrates this situation. We see that 17 "behaves like" 2 in a five-hour clock arithmetic. The number 2 can be determined by dividing 17 by 5 and finding the remainder. In so doing, of course, the

FIGURE 7.4.1

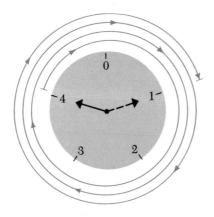

quotient is 3 and the remainder is 2. Actually, the quotient $17 \div 5$ is $3\frac{2}{5}$. However, since the remainder 2 is smaller than the divisor 5, we say that $17 \div 5$ has a "quotient" of 3 with a remainder of 2. This is based on the fact that $17 = 5(3) + 2$. It is undoubtedly a poor decision to use the word quotient in two different but closely related ways. However, we follow a long-standing practice in so doing and no confusion will result if the double usage is understood. Referring once again to the fact that 17 and 2 play a similar role in a five-hour clock, we note that the difference $17 - 2$ is a multiple of 5: $17 - 2 = 3(5)$.

A finite mathematical system, such as a five-hour clock arithmetic, is known as a **modular system** or **congruence** if it is extended in such a way that each integer may be considered equivalent to some element within the finite system. Modular systems have important applications in the realm of rotating objects, which includes wheels, dials, axles, and so on.

In the extension of the five-hour clock arithmetic, 5 is called the **modulus** of the system; the operations and relations on the system are said to be defined *modulo* 5, written (mod 5).

We showed above that 17 is equivalent to 2 in a modulo 5 system. To express this equivalence relation between two integers, we say that they are *congruent* and use the symbol $\equiv$. We read "$17 \equiv 2$ (mod 5)" as "17 is congruent to 2 modulo 5." Now let us formally define the congruence relation.

DEFINITION 7.4.1.   The integer a is **congruent** to the integer b modulo m $(m \in N)$ if and only if $a - b$ is a multiple of m. Symbolically, $a \equiv b$ (mod m) if and only if $a - b = km$ $(k \in J)$.

The above definition could have stated that m must divide $(a - b)$ since this is true if $a - b$ is a multiple of m. It can also be seen that a and b have the same remainder when divided by m. This fact is used by some authors in defining congruence.

Example 1.   Show that $12 \equiv 27$ (mod 5).

Solution: We must show that $12 - 27$ is a multiple of 5. $12 - 27 = -15 = -3(5)$. Therefore, by definition, $12 \equiv 27$ (mod 5).

Example 2.
$19 \equiv 7$ (mod 3) since $19 - 7 = 12 = 4(3)$.
$13 \equiv 37$ (mod 6) since $13 - 37 = -24 = -4(6)$.
$45 \equiv 10$ (mod 7) since $45 - 10 = 35 = 5(7)$.
$13 \not\equiv 8$ (mod 4) since $13 - 8 = 5 \neq k(4)$, where $k \in J$.
$-4 \equiv 14$ (mod 3) since $-4 - 14 = -18 = -6(3)$.
$-19 \equiv -9$ (mod 5) since $-19 - (-9) = -19 + 9 = -10 = -2(5)$.

Example 3. Given $12 \equiv 27$ (mod 5), show that 12 and 27 have the same remainder when divided by 5.

Solution: $12 = 2(5) + 2$ and $27 = 5(5) + 2$. In both cases, 2 is the remainder.

In general, it may be shown that $a \equiv b$ (mod m) if and only if a and b have the same remainder when divided by m. Furthermore, if a divided by m has a remainder r, then $a \equiv r$ (mod m). This is true since $a = km + r$ implies $a - r = km$, and hence by the definition of congruence, $a \equiv r$ (mod m). For example, if we divide 22 by 5, we have $22 = 4(5) + 2$ and $22 \equiv 2$ (mod 5).

Example 4. Determine the smallest positive integer congruent to 44 (mod 6) by dividing and finding the remainder.

Solution: $44 = 7(6) + 2$; therefore, $44 \equiv 2$ (mod 6).

Example 5. Show that $-14 \equiv 2$ (mod 4) by dividing and finding the remainder.

Solution: $-14 = -4(4) + 2$; therefore, $-14 \equiv 2$ (mod 4). Note the temptation to use 3 as a quotient rather than 4; however, this would give us a negative remainder: $-14 = -4(3) + (-2)$.

The congruence relation partitions a set on which it is defined into disjoint subsets or *equivalence classes*. For example, the congruence relation modulo 5 partitions the integers into five equivalence classes that we shall identify as [0], [1], [2], [3], and [4].

$$[0] = \{\ldots, -10, -5, 0, 5, 10, 15, \ldots\}$$
$$[1] = \{\ldots, -9, -4, 1, 6, 11, 16, \ldots\}$$
$$[2] = \{\ldots, -8, -3, 2, 7, 12, 17, \ldots\}$$
$$[3] = \{\ldots, -7, -2, 3, 8, 13, 18, \ldots\}$$
$$[4] = \{\ldots, -6, -1, 4, 9, 14, 19, \ldots\}$$

Note that the difference between any two elements in each equivalence class above is a multiple of 5. Observe also that the members of any particular equivalence class have the same remainder when divided by 5. For example, -6 and 9 are both members of [4], and both have remainders of 4 when divided by 5: $-6 = -2(5) + 4$ and $9 = 1(5) + 4$. Similarly, members of [0] have remainders of 0, members of [1] have remainders of 1, and so on.

Since in a modulo 5 system each integer may be considered equivalent to some element of the finite set $\{0, 1, 2, 3, 4\}$, the addition and multiplication tables of the five-hour clock system may also be used for the modulo 5 system. In constructing the tables, however, our work is facilitated by using the congruence relation. For example, to find the sum $3 + 4$, we have $3 + 4 = 7 = 1(5) + 2$; therefore, $3 + 4 \equiv 2 \pmod 5$. To find a product such as $3 \cdot 2$, we have $3 \cdot 2 = 6 = 1(5) + 1$; therefore, $3 \cdot 2 \equiv 1 \pmod 5$.

It is interesting to learn that in constructing addition and multiplication tables for a modular system, we may use any representative from each of the equivalence classes. For example, in the modulo 5 system we used as representatives the members of $\{0, 1, 2, 3, 4\}$, but we could have used the members of $\{1, 2, 3, 4, 5\}$ or $\{12, 13, 14, 15, 16\}$ or even $\{-10, 6, -3, -7, 19\}$. The only requirement is that we use exactly one representative element from each equivalence class.

Theorem 7.4.1, which is proved below, is a typical result of the congruence relation. The proof is followed by examples illustrating the principles stated in the theorem.

THEOREM 7.4.1.   If $a \equiv b \pmod m$ and $c \equiv d \pmod m$, then:

(i) $a + c \equiv b + d \pmod m$ and
(ii) $ac \equiv bd \pmod m$.

Proof of part (i):
1. $a - b = k_1 m$ and $c - d = k_2 m$ by the definition of congruence.
2. By addition and substitution we have:
   $(a - b) + (c - d) = k_1 m + k_2 m$ or
   $(a + c) - (b + d) = (k_1 + k_2)m = k_3 m$.
3. Therefore, $a + c \equiv b + d \pmod m$ since their difference is a multiple of m.

Proof of part (ii):
1. $a - b = k_1 m$ and $c - d = k_2 m$ by the definition of congruence.
2. Then $a = k_1 m + b$ and $c = k_2 m + d$.
3. By multiplication and substitution we have:
   $ac = k_1 m k_2 m + k_1 md + bk_2 m + bd$ or
   $ac - bd = (k_1 k_2 m + k_1 d + bk_2)m = k_3 m$.
4. Therefore, $ac \equiv bd \pmod m$ since their difference is a multiple of m.

Example 6.   Given $18 \equiv 3 \pmod 5$ and $23 \equiv 8 \pmod 5$, show that $18 + 23 \equiv 3 + 8 \pmod 5$.

Solution: $18 + 23 = 41$ and $3 + 8 = 11$. Since $41 - 11 = 30 = 6(5)$, we see, by the definition of congruence, that $41 \equiv 11 \pmod 5$. It follows, then, that $18 + 23 \equiv 3 + 8 \pmod 5$.

Alternative Solution:

$$18 + 23 = 41 = 8(5) + 1; \text{ therefore, } 41 \equiv 1 \pmod 5.$$
$$3 + 8 = 11 = 2(5) + 1; \text{ therefore, } 11 \equiv 1 \pmod 5.$$

Hence, $18 + 23 \equiv 3 + 8 \pmod 5$.

Example 7. Find the whole number less than the modulus to which the product $7 \cdot 42 \pmod 4$ is congruent.

Solution: $7 \cdot 42 = 294 = 73(4) + 2$; therefore, $7 \cdot 42 \equiv 2 \pmod 4$.

Alternative Solution: $7 \cdot 42 \equiv 7 \cdot 2 \equiv 14 \equiv 2 \pmod 4$.

EXERCISE 7.4

1. Define $a \equiv b \pmod m$.

2. Use the definition of Problem 1 to show that $46 \equiv 22 \pmod 6$.

3. Using division find the smallest whole number remainder when each of the following numbers is divided by 4:
   (a) 13   (b) 15   (c) 2
   (d) -5 (e) -18 (f) 0
   (g) -14 (h) -20 (i) -23

4. (a) Determine the smallest positive integer congruent to 20 modulo 3 by dividing and finding the remainder.
   (b) Check your result in part (a) by using the definition of the congruence relation.

5. Perform the following operations and find the whole number less than the modulus to which each result is congruent:
   (a) $6 \cdot 8 \pmod 5$ (b) $4 + 7 \pmod 3$
   (c) $22 + 33 \pmod 6$ (d) $7 + (9 + 22) \pmod 7$
   (e) $5^3 \pmod 9$ (f) $2 - 7 \pmod 4$
   (g) $16 - 24 \pmod 7$ (h) $-2(-4 + 22) \pmod 5$
   (i) $62 \cdot 35 \pmod 9$ (j) $35 \cdot 24 \pmod{10}$

6. Using the elements in $\{0, 1, 2, 3, 4, 5, 6\}$:
   (a) Construct an addition table modulo 7.
   (b) Construct a multiplication table modulo 7.
   (c) Is the mathematical system a commutative group under addition? Explain.
   (d) Is the mathematical system a commutative group under multiplication? Explain.
   (e) Is the mathematical system a commutative group under multiplication with zero removed?
   (f) Interpret $\sqrt{4} \pmod 7$ and find the two numbers that equal $\sqrt{4} \pmod 7$.

7.5   Congruence and Casting out Nines

For many years "casting out nines" has been used as a partial check for arithmetic computations. With experience it can be applied rapidly and is especially helpful for multiplication and division when large numbers are involved. Although its application is limited, particularly with the widespread use of calculating machines, the theory behind the technique is interesting.

The casting-out-nines check is actually an application of Theorem 7.4.1 in a modulo 9 system. The key to the check by nines is to be able to rapidly find congruences using modulo 9. First note that $1000 - 1 = 999 = k_1(9)$, $100 - 1 = 99 = k_2(9)$, and $10 - 1 = 9 = k_3(9)$. By definition, then, $1000 \equiv 1 \pmod 9$, $100 \equiv 1 \pmod 9$, and $10 \equiv 1 \pmod 9$. It can be shown that, in general, any power of 10 is congruent to 1 modulo 9. Furthermore, it can be shown that any number is congruent to the sum of its digits modulo 9. For example, $8234 \equiv 8(1000) + 2(100) + 3(10) + 4 \equiv 8(1) + 2(1) + 3(1) + 4 \equiv 8 + 2 + 3 + 4 \equiv 17 \pmod 9$. Similarly, $17 \equiv 1(10) + 7 \equiv 1(1) + 7 \equiv 1 + 7 \equiv 8 \pmod 9$. Therefore, $8234 \equiv 8 \pmod 9$. The algorithm may be shortened to $8234 \equiv 8 + 2 + 3 + 4 \equiv 17 \equiv 1 + 7 \equiv 8 \pmod 9$.

Using the concepts of the previous paragraph along with Theorem 7.4.1, we can partially check addition, multiplication, and division, as shown in the following examples. We have only a *partial check* by "casting out nines" because any number congruent to the correct result modulo 9 would "check," even if it were not a correct result. The chance of having an incorrect result that "checks" is remote but certainly possible.

Example 1.   Find the sum $248 + 324 + 672$ and check by casting out nines.

Solution:

$$248 \equiv 2 + 4 + 8 \equiv 14 \equiv 5 \pmod 9$$
$$324 \equiv 3 + 2 + 4 \equiv 9 \equiv 0 \pmod 9$$
$$672 \equiv 6 + 7 + 2 \equiv 15 \equiv 6 \pmod 9$$
$$11 \equiv \boxed{2} \pmod 9$$

↕ check number

$$1244 \equiv 1 + 2 + 4 + 4 \equiv 11 \equiv \boxed{2} \pmod 9$$

We can shorten the above work in several ways. For example, in finding $672 \equiv 6 \pmod 9$, we may add 6 and 7 and obtain 13; then add the digits of 13 and obtain 4; and finally add 4 to the remaining digit 2 and obtain 6. By this process we may simply write $672 \equiv 6$ (mod 9). All the manipulations may be performed mentally in rapid succession so that nothing is written except the final result.

Example 2. Find the product 84×428 and check by casting out nines.

Solution:

$$
\begin{array}{r}
428 \equiv 14 \equiv 5 \ (\text{mod } 9) \\
84 \equiv 12 \equiv 3 \ (\text{mod } 9) \\
\hline
1712 \qquad 15 \equiv \boxed{6} \ (\text{mod } 9)
\end{array}
$$

3424   check number

$35952 \equiv 24 \equiv \boxed{6}$ (mod 9)

Example 3. Find the quotient and remainder for $24{,}532 \div 824$ and check by casting out nines.

Solution:

$$
\begin{array}{r}
29 \\
824\overline{)24532} \\
1648 \\
\hline
8052 \\
7416 \\
\hline
636 \quad (\text{remainder})
\end{array}
$$

In checking, we must show that $24{,}532 \equiv 824(29) + 636$ (mod 9).

$24{,}532 \equiv 2 + 4 + 5 + 3 + 2 \equiv 16 \equiv \boxed{7}$ (mod 9)

check number

$824(29) + 636 \equiv 5(2) + 6 \equiv 10 + 6 \equiv 16 \equiv \boxed{7}$ (mod 9)

EXERCISE 7.5

1. Cast out nines from the following numbers:
   (a) 23   (b) 264
   (c) 23,425   (d) 62,397
   (e) 823,429   (f) 6,234,425

2. Add the following and check by casting out nines:
   (a) $534 + 928$ (b) $325 + 6243$
   (c) $62{,}345 + 82{,}347$ (d) $234 + 9235 + 17{,}238$

3. Multiply the following and check by casting out nines:
   (a) $34 \cdot 523$ (b) $22 \cdot 492$
   (c) $27 \cdot 32,446$ (d) $444 \cdot 6724$

4. Divide the following and check by casting out nines:
   (a) $624 \div 34$ (b) $8429 \div 67$
   (c) $26,000 \div 14$ (d) $82,925 \div 643$

5. Explain how casting out nines can be used to check subtraction.

REVIEW EXERCISES

1. Construct an addition table for a three-hour clock arithmetic using elements of $\{0, 1, 2\}$.

2. Refer to the three-hour clock addition table of Problem 1 and answer the following:
   (a) What is the identity element for addition?
   (b) What is the additive inverse of each element?
   (c) Is the system a commutative group under addition? Explain.

3. Construct a multiplication table for a three-hour clock arithmetic using elements of $\{0, 1, 2\}$.

4. Refer to the three-hour clock multiplication table of Problem 3 and answer the following:
   (a) What is the identity element for multiplication?
   (b) What is the multiplicative inverse (if any) of each element?
   (c) Is the system a commutative group under multiplication? Explain.
   (d) Is a three-hour clock arithmetic a field? Explain.

5. Make a multiplication table for a six-hour clock arithmetic using elements of $\{0, 1, 2, 3, 4, 5\}$.

6. Refer to the six-hour clock multiplication table of Problem 5 and answer the following:
   (a) What is the multiplicative inverse (if any) of each element?
   (b) Is the system a commutative group under multiplication? Explain.
   (c) Is a six-hour clock arithmetic a field? Explain.

7. If a missile goes into orbit about the earth at 10:00 A.M. and stays in orbit for 1000 hours, what time will it land?

8. Perform the following operations and find the whole number less than the modulus to which each result is congruent:
   (a) $7 \cdot 4 \pmod 5$ (b) $8 + 6 \pmod 3$
   (c) $24 + 33 \pmod 6$ (d) $4^3 \pmod 7$
   (e) $18 - 13 \pmod 5$ (f) $15 - 25 \pmod 3$

9. Find the product 38×653 and check by casting out nines.

10. Find the quotient and remainder for $24,436 \div 843$ and check by casting out nines.

11. Using a five-hour clock arithmetic:
    (a) What would $1 - 2$ equal?
    (b) What would $3 - 4$ equal?
    (c) What would $1 - 3$ equal?
12. Using a five-hour clock arithmetic:
    (a) What would $4 \div 3$ equal?
    (b) What would $3 \div 2$ equal?
    (c) What would $1 \div 3$ equal?

13. Name the additive inverse of each element in a five-hour clock arithmetic.

14. Name the multiplicative inverse of each element (except zero) in a five-hour clock arithmetic.

Victor Vasarely, *Orion Blanc*
1964. Oil on canvas. 82½ x 78½".
Courtesy, the Cleveland Museum of Art.

8 PROBABILITY

8.1 Introduction

Early Greek philosophers, such as Aristotle and his colleagues, are known to have discussed concepts of luck, chance, and choice. Although these topics are related to the concept of probability, there was little mathematical treatment of probability until the seventeenth century, and then it developed in connection with gambling. This is not surprising since the concept of "something for nothing" seems to have universal appeal. Although probability theory is still very much a part of what Nevadans call the "gaming industry," applications of probability theory now permeate many facets of modern culture. Today, concepts of probability are applied in such diverse fields as mortality and insurance, biology, theoretical physics, educational measurements, and sociology. However, as you will see in this chapter, games of chance still provide excellent vehicles for introducing and illustrating elementary concepts of probability.

Although Pierre de Fermat is best known for his work in the theory of numbers, he and Blaise Pascal (1623–1662) are considered pioneers in the development of probability theory. An initial factor in stimulating their interest in probability was a question posed for Pascal by Chevalier de Mére, a gambler and member of the French aristocracy. Mére's question concerned the problem of dividing the stakes in a game of chance if two players of equal skill were forced to end the game prior to its completion. It seemed obvious that if a game was prematurely terminated, the stakes should be divided according to each player's chance of winning had the game been completed. Pascal corresponded with Fermat concerning the problem and although each one solved the problem differently, they arrived at the same conclusion concerning the manner in which the stakes should be divided.

We have already recounted several of the more interesting aspects of Fermat's life (Chapter 6) and need not repeat them here. However, it seems appropriate to comment briefly on Pascal's life. Pascal was a child prodigy in mathematics and had proved many theorems in geometry by the time he was 12 years old. Before the age of 20 he had done important original work concerning circles, parabolas, ellipses, and so on, and had invented a calculating machine to assist his father in handling quantities of statistical data as they related to his job as an administrator. Had Pascal continued with his work in mathematics, he no doubt would have become one of the most prolific mathematicians in history. However, Pascal was in poor health much of his life and was, at least by modern standards, somewhat of a religious mystic. In 1654 Pascal was nearly killed by a runaway horse. The fact that he survived the accident was interpreted by Pascal as a sign that he should devote the rest of his life to religious contemplation, which with rare exceptions he did.

Inspired by Fermat and Pascal, mathematicians continued to develop probability theory. Among the significant contributors are Christian Huygens (1629–1695), who published a treatise on the theory of probability, and Jacques Bernoulli (1654–1705), who suggested applications of probability to civil, moral, and economic affairs. Others, such as Abraham De Moivre, Euler, and Gauss, may be cited for their contributions. Pierre-Simon Laplace (1749–1827) also made notable contributions to probability theory. Although Laplace came from a poor peasant home he managed to divorce himself from his humble origins, perhaps to the point of snobbery. By virtue of his great mathematical talent, high self-confidence, and skill in politics, Laplace became a Professor of Mathematics at the Military School of Paris, obtained full membership in The Academy of Science, was awarded many honors by Napoleon and by Louis XVIII, and finally retired at his country estate near Paris. By shifting his political allegiance to whoever happened to be in power, Laplace became known as a political opportunist. For example, he supported Napoleon and as a reward was given the title of Count; however, shortly thereafter Laplace signed a decree banishing Napoleon and was among the first to support the returning king. Nevertheless, during his long survival in the world of French politics, Laplace managed to make important discoveries in both probability theory and astronomy. In fact, he has been called the "Newton of France."

Over a period of 300–400 years, probability theory has grown from a few suggestions for gamblers to a deductive mathematical system. This fact is well exhibited in a 1933 publication, *Foundations of the Theory of Probability*, by the Russian mathematician A. N. Kolmogorov. The list of applications of probability theory continues to grow at a rapid pace.

8.2 Counting Techniques and Factorial Notation

Before we define probability, it is appropriate to discuss a few preliminary topics, such as counting techniques, factorial notation, permutations, and combinations. In this section, counting techniques and factorial notation are considered. We begin with a few comments on counting techniques.

If we wish to determine the number of elements in a given finite set, we can count them by ordinary counting methods. However, more sophisticated methods may be used for counting in certain special situations. For example, if we know there are 18 boys and 12 girls in a class of students, we can find the number of students in the entire class simply by finding the sum $18 + 12 = 30$. We might think of this as a quick method of counting since we have determined the cardinal number of the given set of students. Note that the set of boys B and the set of girls G are disjoint sets and that we found the number of elements in the union of the two sets: $n(B \cup G) = n(B) + n(G) = 18 + 12 = 30$.

Referring to the class of students mentioned in the previous paragraph, let us consider a different kind of problem. Suppose we wish to elect a committee of two students to plan a party for the class. If the committee is to be composed of a boy and a girl, in how many ways may the committee be chosen? Since each of the 18 boys may be paired with each of the 12 girls, the number of possible committees is $18 \times 12 = 216$. Note that if B is the set of boys and G is the set of girls, we could find the number of committees by naming and counting all of the ordered pairs of elements in the product set $B \times G$; but this would hardly be exciting and is unnecessary since $n(B \times G) = n(B) \cdot n(G) = 18 \times 12 = 216$. Furthermore, since it makes no difference whether we choose a boy first or a girl first, we could equally well find the number of possible committees by determining the number of elements in $G \times B$: $n(G \times B) = n(G) \cdot n(B) = 12 \times 18 = 216$.

Example 1. If there are 4 acceptable routes from Las Vegas to Reno and 3 routes from Reno to San Francisco, how many different routes may be taken in getting from Las Vegas to San Francisco via Reno? Illustrate with a diagram.

Solution: Since each of the 4 routes from Las Vegas to Reno may be paired with each of the 3 routes from Reno to San Francisco, the number of possible routes is $4 \times 3 = 12$. Referring to Figure 8.2.1 we see that the 12 possible routes may be indicated by: *ax, ay, az, bx, by, bz, cx, cy, cz, dx, dy,* and *dz.*

FIGURE 8.2.1

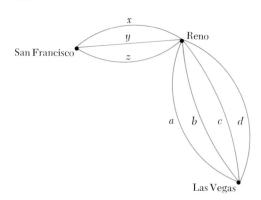

Now let us consider an example that is slightly more complicated but similar to Example 1. Suppose a person has 3 different suits and 4 different neckties. In how many ways can he dress? With each of the 3 suits he can wear 4 different neckties; hence he can dress in $3 \times 4 = 12$ different ways. Suppose he also has 5 different pairs of shoes. Then each pair of shoes can be worn with each of the previous 12 ways of dressing, making $5 \times 12 = 60$ different ways of dressing; or the number of ways of dressing is $3 \times 4 \times 5$. In general, if there are a choices for doing one thing, b choices for doing another, c choices for another, and so forth, then *together* they may be done in $a \times b \times c \times \ldots$ different ways. This principle is often called the fundamental counting principle.

Example 2. In how many ways may we seat a class of 30 students in a room with 30 chairs?

Solution: The first student to be considered may sit in any of 30 chairs, the next in any of 29, the next in any of 28, and so on. Hence, using the fundamental counting principle, we conclude that there are $30 \times 29 \times 28 \times \ldots \times 2 \times 1$ different possible seating arrangements.

We may abbreviate the expression $30 \times 29 \times 28 \times \ldots \times 2 \times 1$ in Example 2 by writing 30!, which is read "thirty factorial" and indicates the product of all the integral factors from 30 down to 1. Similarly,

$$1! = 1$$
$$2! = 2 \times 1 = 2$$
$$3! = 3 \times 2 \times 1 = 6$$
$$4! = 4 \times 3 \times 2 \times 1 = 24$$
$$5! = 5 \times 4 \times 3 \times 2 \times 1 = 120$$

If $n \geq 1$, we define

$$n! = n(n - 1)(n - 2) \cdots (2)(1).$$

We will also find it advantageous to define $0! = 1$. It is interesting to observe how rapidly $n!$ increases as n increases. For example, if 12! ordinary postage stamps were placed end to end, they would more than reach from Los Angeles to New York City and back again. With 20! postage stamps we could cover the entire surface of the earth twice. It has been computed that if the earth were hollow, 30! drops of water would more than fill it 20,000 times. Sir Arthur Eddington (1882–1944), an English astronomer and physicist, claimed the number of electrons in the entire universe to be 136×2^{256} or about 1.57×10^{79}. Using factorial notation, we find that 59! is more than eight times as large as this number.

One more example using the fundamental counting principle should be sufficient to illustrate its value as well as the variety of situations in which it may be applied.

Example 3. If a married couple has 4 children and plans to take a trip, how many different sets of their children could be selected to go with them (including the empty set)?

Solution: There are exactly two choices for each child—either he is selected or he is not selected for the trip. Therefore, by the fundamental counting principle, the total number of ways of selecting a set of children for the trip is $2 \times 2 \times 2 \times 2 = 2^4 = 16$.

The solution to Example 3 above is, essentially, a matter of finding the total number of subsets of a given set. In general, since each element in a set may be either accepted or rejected in composing a subset, there are 2^n possible subsets for a set containing n elements. Subsets of a given set, of course, include both the set itself and the empty set.

EXERCISE 8.2

1. If each of 3 models of a refrigerator is available in 5 different colors, how many different orders may be placed for a refrigerator?

2. Evaluate:
   (a) 0!   (b) 1!   (c) (3!)(5!)   (d) $\frac{5!}{4!}$ (e) $\frac{26}{2!}$

3. How many different two-digit numerals may be formed from the set
   $S = \{2, 4, 6, 8\}$:
   (a) If no digit may be repeated?
   (b) If any of the digits may be repeated?

4. A college girl has 6 sweaters, 5 skirts, and 2 pairs of shoes. If she wears
   one of each, in how many ways can she dress?

5. In buying an automobile there are 8 optional extras, such as radio, air
   conditioner, and power steering. How many different sales orders are
   possible?

6. If a true-false test has 10 statements and each statement is marked either
   true or false, in how many ways may the entire test be answered?

7. In how many ways may a president, vice-president, secretary, and
   treasurer be elected from a class of 25 students, if no person may hold
   two offices?

8. (a) In how many ways may 5 people be seated in 5 chairs?
   (b) In how many ways may 60 people be seated in 60 chairs?

9. Blind persons read braille. A braille "cell" consists of 3 locations in
   each of 2 columns or 6 locations in all. Each location may or may not
   have a raised dot, which can be read by touch. If at least one dot is
   necessary for a symbol, how many different letters, numerals, and so
   on may be represented by the 6 dots of a cell?

10. If any sequence of letters is considered to be a "word," find the number
    of different four-letter words that may be formed from our alphabet:
    (a) If no letter may be repeated.
    (b) If any letter may be repeated.
    (c) If any of the letters may be repeated, but the first letter of the word
        must be x, y, or z.

11. Assume there are 5 routes from Philadelphia to Atlanta and 5 routes
    from Atlanta to Miami. In how many ways may a person make a round
    trip from Philadelphia to Miami, going through Atlanta in both direc-
    tions, without driving on the same route twice?

8.3 Permutations

In Example 2 of the previous section we discussed the various ways of
seating 30 students in 30 chairs and showed the total number of arrangements to be 30!. Each of these 30! different ways of ordering the students
is called a *permutation*. In general, each *ordered* set that may be obtained
from a given set is called a **permutation**. We may permute or change the
order of the elements of a given set by using all the elements of the
set or by using only part of the elements. For example, the permutations

of the elements of set $S = \{a, b, c\}$ using all the elements are abc, acb, bac, bca, cab, and cba; using only two of the three elements at a time, the permutations are ab, ac, ba, bc, ca, and cb; and using only one of the elements at a time, the permutations are a, b, and c.

Various types of symbolism are used to represent the *number of permutations* of a set of n elements taken r at a time; some of them are: $P(n, r)$, P_r^n, and $_nP_r$. We will use the symbol $P(n, r)$ to denote the number of permutations of n elements taken r at a time. Obviously, $P(n, r)$ is such that $r \leq n$. If all the elements of the set are used for each permutation, we have $P(n, n)$. Referring to the $S = \{a, b, c\}$ mentioned in the previous paragraph, and using the fundamental counting principle, we see that:

$$P(3, 3) = 3 \cdot 2 \cdot 1 = 6$$
$$P(3, 2) = 3 \cdot 2 = 6$$
$$P(3, 1) = 3$$

Example 1. If a different person must be selected for each position, in how many ways may we select a president, vice-president, treasurer, and secretary from a class of 30 students?

Solution: We may think of the first person in a set of four persons as the president, the next as the vice-president, the next as the treasurer, and the last as the secretary. Then each different ordering of four persons chosen from among the 30 students represents a different set of class officers. Hence the permutations of 30 elements taken 4 at a time will give us the desired result. Since the first of the 4 elements may be chosen 30 different ways, the next 29, the next 28, and the last 27, the desired result is:

$$P(30, 4) = 30 \cdot 29 \cdot 28 \cdot 27 = 657{,}720.$$

Note that in finding the value of $P(30, 4)$ we use exactly four factors and that, in general, $P(n, r)$ can be found by using r factors. Applying the fundamental counting principle we have

$$P(n, r) = n(n - 1)(n - 2) \cdots \text{ to } r \text{ factors.}$$

An alternative way of expressing $P(n, r)$ is

$$P(n, r) = n(n - 1)(n - 2) \cdots (n - r + 1).$$

If the right-hand member of the above equation is multiplied by $(n - r)!/(n - r)!$, we obtain the compact equation

$$P(n, r) = \frac{n!}{(n - r)!}.$$

In the event that $r = n$ we have

$$P(n, n) = n!.$$

The meaning of $P(n, 0)$ is somewhat vague; however, as we will see in Section 8.8, it is advantageous to define $P(n, 0) = 1$.

Example 2. If any ordered set of letters is considered to be a "word," how many different five-letter words may be spelled using our alphabet:

    (a) If a letter may not be used more than once.
        Solution: $P(26, 5) = 26 \cdot 25 \cdot 24 \cdot 23 \cdot 22 = 7{,}893{,}600$.
    (b) If a letter may be repeated as often as desired.
        Solution: Using the fundamental counting principle we see the number of possibilities is $26 \cdot 26 \cdot 26 \cdot 26 \cdot 26 = 26^5 = 11{,}881{,}376$.

Example 3. In how many ways can the members of $\{A, B, C, 1, 2, 3, 4, 5\}$ be ordered if the letters must come first?

Solution: The letters may be arranged in 3! different ways and the digits in 5! different ways. Using the fundamental counting principle we obtain a final result of $3!5! = 6(120) = 720$.

    In some sets of elements there may be certain members that are indistinguishable from each other. Example 4 illustrates how to find the number of permutations in this kind of situation.

Example 4. In how many ways may the letters of the word *PARALLEL* be arranged to form a new "word."

Solution: If each of the 8 letters of *PARALLEL* were different, there would be $P(8, 8) = 8!$ different possible words. However, the 3 *L*'s are indistinguishable from each other and may be permuted in 3! different ways. As a result, each of the 8! arrangements of the letters of *PARALLEL* that would otherwise spell a new word will be repeated 3! times. To avoid counting repetitions resulting from the 3 indistinguishable *L*'s in *PARALLEL*, we may divide 8! by 3!. Similarly, the 2 *A*'s in *PARALLEL* may be arranged in 2! different ways and we must divide by 2! to avoid counting repetitions resulting from the 2 indistinguishable *A*'s. Hence the total number of words that may be formed from the letters of the word *PARALLEL* is

$$\frac{P(8, 8)}{3!\, 2!} = \frac{8!}{3!\, 2!} = 3360.$$

    In generalizing the example above, we see that if a set of n elements has k_1 indistinguishable elements of one kind, k_2 of another kind, and so on for r kinds of elements, then the number of permutations of the set of n elements is

$$\frac{n!}{k_1! \, k_2! \, \cdots \, k_r!}.$$

EXERCISE 8.3

1. Evaluate each of the following:
   (a) $P(5, 0)$ (b) $P(5, 1)$ (c) $P(5, 2)$
   (d) $P(5, 3)$ (e) $P(5, 4)$ (f) $P(5, 5)$

2. In how many ways may 6 books be arranged in a row on a shelf?

3. (a) In how many different ways may 3 people be seated in 3 chairs?
   (b) In how many different ways may 3 people be seated in 7 chairs?

4. How many different 3-digit license-tag numerals may be formed using members of $\{1, 2, 3, 4, 5, 6\}$:
   (a) If no digit may be repeated?
   (b) If any digit may be repeated?

5. Telephone numbers with area codes have ten digits and are of the form xxx-xxx-xxxx. With the restrictions that the second digit must be 0 or 1 and the fourth digit may not be 0, how many different telephone numbers can be formed?

6. If women must go first, in how many ways may 6 women and 4 men enter a lifeboat?

7. How many different numbers may be indicated by rearranging the 7 digits of 9,666,555?

8. If marbles of the same color are indistinguishable, how many different orderings of a row of 2 red, 3 white, and 4 blue marbles may be recognized?

9. If 10 different flags are available and a signal consists of 3 flags flown in a vertical column, how many different signals are possible?

10. In how many ways can 4 mathematics books and 3 history books be arranged on a shelf:
    (a) With no restrictions?
    (b) If all of the mathematics books are on the left of the history books?

11. In how many ways may 4 men and 3 women stand in a row to be photographed:
    (a) If all members of the same sex stand side by side?
    (b) If all of the men stand side by side?

8.4   Combinations

Suppose we wish to appoint a committee of 4 persons from a class of 30 students. We know that $P(30, 4)$ is the number of different *ordered* sets of 4 students each that may be selected from among 30 students. However, since the ordering of the students on the committee has no significance, our problem is to determine the number of four-element *unordered* subsets that may be constructed from a set of 30 elements. Since any four-element set may be ordered in 4! different ways, $P(30, 4)$ is 4! times too large. Hence if we divide $P(30, 4)$ by 4!, the result will be the number of *unordered* subsets of 30 elements taken 4 at a time. The number of unordered subsets of 30 elements taken 4 at a time is also called the number of *combinations* of 30 elements taken 4 at a time and may be indicated as $C(30, 4)$. From the above discussion, we see that we can find the number of combinations of 30 elements taken 4 at a time as follows:

$$C(30, 4) = \frac{P(30, 4)}{4!} = \frac{30 \cdot 29 \cdot 28 \cdot 27}{4 \cdot 3 \cdot 2 \cdot 1} = 27{,}405.$$

In general, each *unordered* r-element subset of a given n-element set $(r \le n)$ is called a **combination**. The *number of combinations* of n elements taken r at a time is indicated by $C(n, r)$. A general equation relating combinations to permutations is

$$C(n, r) = \frac{P(n, r)}{r!} = \frac{n(n - 1)(n - 2) \cdots (n - r + 1)}{r!}.$$

Multiplying the right member by $(n - r)!/(n - r)!$ we have alternatively

$$C(n, r) = \frac{n!}{r! \, (n - r)!}.$$

Example 1. From among 20 boys who play basketball, in how many different ways can a team of 5 players be selected?

Solution: Since order of selection is of no consequence, we use combinations:

$$C(20, 5) = \frac{P(20, 5)}{5!} = \frac{20 \cdot 19 \cdot 18 \cdot 17 \cdot 16}{5 \cdot 4 \cdot 3 \cdot 2 \cdot 1} = 15{,}504.$$

As compared with Example 1, a slightly more complex problem is discussed in Example 2. Often, in more difficult problems, alternative but equivalent solutions may be found.

Example 2. From among 20 boys who play basketball, in how many different ways can a team of 5 players be selected if one of the players is to be designated as captain?

Solution: Without designating a captain the result would be $C(20, 5)$. However, 5 different captains may be chosen in each different set of players.

Therefore, using the fundamental counting principle, we must multiply $C(20, 5)$ by 5 to obtain the total number of possibilities. The result is

$$5[C(20, 5)] = 5\left[\frac{P(20, 5)}{5!}\right] = 5\left[\frac{20 \cdot 19 \cdot 18 \cdot 17 \cdot 16}{5 \cdot 4 \cdot 3 \cdot 2 \cdot 1}\right] = 77{,}520.$$

Alternative Solution: A captain may be chosen from any of the 20 players. The remaining 4 players may be chosen in $C(19, 4)$ different ways. Using the fundamental counting principle, the total number of different teams that may be formed is

$$20[C(19, 4)] = 20\left[\frac{P(19, 4)}{4!}\right]$$

$$= 20\left[\frac{19 \cdot 18 \cdot 17 \cdot 16}{4 \cdot 3 \cdot 2 \cdot 1}\right]$$

$$= 20(3876)$$

$$= 77{,}520$$

Example 3. How many different 6-card hands may be dealt from a deck of 52 playing cards?

Solution: Since we are not concerned with the order in which each hand is dealt, our problem concerns the number of combinations of 52 elements taken 6 at a time. Hence we have

$$C(52, 6) = \frac{P(52, 6)}{6!} = \frac{52 \cdot 51 \cdot 50 \cdot 49 \cdot 48 \cdot 47}{6 \cdot 5 \cdot 4 \cdot 3 \cdot 2 \cdot 1} = 20{,}358{,}520.$$

Example 4. How many different 6-card hands containing 3 aces may be dealt from a 52-card deck?

Solution: The 3 aces may be selected from among the 4 aces of the deck in $C(4, 3)$ different ways. The remaining 3 cards of the hand must be selected from among the 48 cards that are not aces; hence they may be selected in $C(48, 3)$ different ways. Using the fundamental counting principle, the total number of possible hands is

$$[C(4, 3)] \cdot [C(48, 3)] = \frac{P(4, 3)}{3!} \cdot \frac{P(48, 3)}{3!}$$

$$= \frac{4 \cdot 3 \cdot 2}{3 \cdot 2 \cdot 1} \cdot \frac{48 \cdot 47 \cdot 46}{3 \cdot 2 \cdot 1}$$

$$= 4(17{,}296)$$

$$= 69{,}184$$

Example 5. If a married couple has 4 children and plans to take a trip, how many different sets of their children could be selected to go with them (including the empty set)?

Solution: This problem is a restatement of Example 3 of Section 8.2. Since there are exactly two choices for each child (he is either selected or not selected), the application of the fundamental counting principle gives the result $2^4 = 16$. This result can be verified by an interesting (but much longer) solution involving the use of combinations. Since the couple may take a set of 0, 1, 2, 3, or 4 children on the trip, the total number of possible unordered sets is

$$C(4, 0) + C(4, 1) + C(4, 2) + C(4, 3) + C(4, 4)$$
$$= 1 + 4 + 6 + 4 + 1 = 16.$$

EXERCISE 8.4

1. Recalling that we defined $P(n, 0) = 1$ and $0! = 1$, show that $C(n, 0) = 1$.

2. Evaluate:
    (a) $C(5, 0)$ (b) $C(5, 1)$ (c) $C(5, 2)$
    (d) $C(5, 3)$ (e) $C(5, 4)$ (f) $C(5, 5)$

3. If a menu lists 10 items, in how many ways may a customer place an order for 5 items?

4. Given 10 points such that no 3 are on the same line, how many different triangles may be drawn by choosing their vertices from among the 10 points?

5. If a penny, nickel, dime, quarter, and half-dollar are dropped on the floor, in how many different ways may they land such that there are
    (a) 3 heads up?
    (b) 4 heads up?
    (c) 5 heads up?

6. There are 10 items to be given away in a contest. The first winner may select any 3 of the items as his prize. In how many different ways may this be done?

7. How many different ways may a committee of 5 be selected from among the 100 U. S. Senators if exactly 1 of the 2 Senators from New York is a member of the committee?

8. A deck of 52 cards contains 13 spades. In how many ways may a hand of 5 cards be dealt if it is to contain:
    (a) Exactly 3 spades?
    (b) At least 3 spades?
    (c) At most 1 spade?

8.5   The Definition of Probability

The topics of probability and statistics are closely related and supplement each other in many ways. Applications of probability and statistics to gambling are familiar to many people and have been studied for years. More recently, applications to such diverse fields as physics, biology, economics, psychology, and medicine have proved most rewarding.

We can introduce the concept of probability by referring to some relatively simple situations. Suppose, for example, we know that there are 50 black marbles and 50 white marbles in a jar. *Probability* tells us that if the marbles are thoroughly mixed and we take a random sample of 10 marbles, we are likely to get about an equal number of black and white marbles. Suppose, however, that we have a jar containing 100 marbles of unknown color. If we select a random sample of 10 marbles from the jar and there are 5 black marbles and 5 white marbles in the sample, *statistical inference* enables us to make a prediction concerning how many of the total of 100 marbles are likely to be black and how many are likely to be white.

Probability is a measure of "chance." Let us consider events with equally likely outcomes. For example, assume that in a deck of 52 well-shuffled cards, each card is equally likely to be selected. What is the chance that if a card is selected, it will be the ace of spades? Since there is only one ace of spades, we have 1 chance out of 52, and we say that the probability of drawing the ace of spades is $\frac{1}{52}$. What is the chance that a card selected at random will be a heart? Since there are 13 hearts, we have 13 chances out of 52; or, since $\frac{13}{52} = \frac{1}{4}$, we have 1 chance out of 4, and the probability of selecting a heart is $\frac{13}{52}$ or $\frac{1}{4}$. With this rather informal introduction, let us define the probability of an event.

DEFINITION 8.5.1. If in a given situation there are n equally likely outcomes, and if a of these outcomes correspond to event A, then the **probability** of A is $\frac{a}{n}$. Symbolically, $P(A) = \frac{a}{n}$.

If $\overline{A}$ corresponds to *not-A*, it follows from Definition 8.5.1 that

$$P(\overline{A}) = \frac{n-a}{n} = 1 - \frac{a}{n} = 1 - P(A).$$

Example 1. If there are 25 names in a hat, 15 girl's names and 10 boy's names, what is the probability that a name drawn at random will be:
(a) A girl's name?
Solution: $P(G) = \frac{15}{25} = \frac{3}{5}$.

(b) A boy's name?

Solution: $P(B) = \frac{10}{25} = \frac{2}{5}$.

(c) Not a girl's name?

Solution: $P(\overline{G}) = 1 - P(G) = 1 - \frac{3}{5} = \frac{2}{5}$.

(d) Not a boy's name?

Solution: $P(\overline{B}) = 1 - P(B) = 1 - \frac{2}{5} = \frac{3}{5}$.

Example 2. If an ordinary die (plural, *dice*) having the numbers 1, 2, 3, 4, 5, and 6 on its faces is rolled, what is the probability that:

(a) The number 4 will be face up?

Solution: $P(4) = \frac{1}{6}$.

(b) An even number will be face up?

Solution: $P(\text{even number}) = \frac{3}{6} = \frac{1}{2}$.

(c) A 7 will be face up?

Solution: $P(7) = \frac{0}{6} = 0$.

(d) A number from 1 to 6 will be face up?
Solution: $P(1, 2, 3, 4, 5, \text{ or } 6) = \frac{6}{6} = 1$.

The solutions to parts (c) and (d) of the above example illustrate that the probability of an event that cannot happen is 0, while the probability of an event that is certain to happen is 1. Hence $0 \leq P(A) \leq 1$.

A set of possible outcomes or elements in a given "experiment" is frequently referred to as a **sample space** of the experiment. The *sample space* used for a given experiment may also be called the *universal set* for the experiment. In Example 1 of this section, the sample space that we used was the set S of names in a hat, and the number of elements in the sample space was $n(S) = 25$. In Example 2 the sample space was the set of numbers on the faces of a die or $S = \{1, 2, 3, 4, 5, 6\}$; the number of elements in this sample space is, of course, $n(S) = 6$.

In general, the experiments which we discuss will be such that a sample space may be chosen with elements that are *equally likely* events. For example, we will not roll "loaded" dice or flip weighted coins. We will not deal from "stacked" decks of cards, but from decks where each card is equally likely to be chosen.

Example 3. Assume a pair of dice is rolled (or one die is rolled twice):

(a) In how many ways may they come to rest?
Solution: Since one of the dice may have any of 6 faces up and the other die may also have any of 6 faces up, the sample space of the experiment would have $6 \times 6 = 36$ elements.

(b) Name the elements of the sample space.

Solution: $\{(1, 1), (1, 2), (1, 3), (1, 4), (1, 5), (1, 6), (2, 1), (2, 2), (2, 3), (2, 4), (2, 5), (2, 6), (3, 1), (3, 2), (3, 3), (3, 4), (3, 5), (3, 6), (4, 1), (4, 2), (4, 3), (4, 4), (4, 5), (4, 6), (5, 1), (5, 2), (5, 3), (5, 4), (5, 5), (5, 6), (6, 1), (6, 2), (6, 3), (6, 4), (6, 5), (6, 6)\}$.

(c) What is the probability that the sum of the face numbers on the dice will be 4?

Solution: Since the sample space has 36 elements and there are 3 ways of rolling a sum of 4, $(1, 3)$, $(2, 2)$, and $(3, 1)$, we see that

$$P(\text{rolling a } 4) = \tfrac{3}{36} = \tfrac{1}{12}.$$

Example 4. If a committee of 3 is chosen at random from among 5 girls and 4 boys, what is the probability the committee will:

(a) Contain all girls?

Solution: Since the order in which committee members are chosen makes no difference, we are concerned with combinations rather than permutations. Hence the number of ways all girls may be chosen for the committee is $C(5, 3)$, and the number of ways any committee of 3 members may be chosen from the entire group of 9 people is $C(9, 3)$. Therefore,

$$P(3 \text{ girls}) = \frac{C(5, 3)}{C(9, 3)} = \frac{\dfrac{5 \cdot 4 \cdot 3}{3!}}{\dfrac{9 \cdot 8 \cdot 7}{3!}} = \frac{5 \cdot 4 \cdot 3}{9 \cdot 8 \cdot 7} = \frac{5}{42}.$$

(b) Contain 2 girls and 1 boy?

Solution: As in part (a), order is irrelevant and we use combinations. Two girls may be chosen in $C(5, 2)$ ways, while one boy may be chosen in $C(4, 1)$ ways. Together they may be chosen in $[C(5, 2)] \cdot [C(4, 1)]$ ways. Any committee of 3 members may be chosen in $C(9, 3)$ ways. Therefore,

$$P(2 \text{ girls and 1 boy}) = \frac{[C(5, 2)] \cdot [C(4, 1)]}{C(9, 3)}$$

$$= \frac{\dfrac{5 \cdot 4}{2!} \cdot \dfrac{4}{1!}}{\dfrac{9 \cdot 8 \cdot 7}{3!}}$$

$$= \frac{10}{21}$$

Example 5. A typical Las Vegas roulette wheel has numbers from 1 to 36 as well as the elements 0 and 00.

(a) If you bet that some odd number will be selected and the casino wins on any even number including 0 as well as 00, what is the probability you will win?

Solution: $P(\text{you win}) = \frac{18}{38} = \frac{9}{19}$.

(b) What is the probability that the casino will win?
Solution: $P(\text{casino wins}) = \frac{20}{38} = \frac{10}{19}$.

Alternative Solution:

$$P(\text{casino wins}) = 1 - P(\text{you win})$$
$$= 1 - \frac{9}{19} = \frac{10}{19}$$

Example 6. In roulette, what is the probability that one of the numbers 3, 6, 9, 12, or 15 will be a winning number?

Solution: Since any of 5 numbers may be designated to win and there are 38 elements in our sample space, the probability of a winning number is $\frac{5}{38}$.

8.6 Odds

The term *odds* is frequently used in speaking of probabilities. The relative chance that an event A will occur as compared with $\bar{A}$ (not-A) is expressed as the *odds in favor of A*.

DEFINITION 8.6.1. If in a given situation there are n equally likely outcomes, and if a of these outcomes correspond to event A, then the **odds** favoring A are

$$\frac{P(A)}{P(\bar{A})} = \frac{\dfrac{a}{n}}{\dfrac{n-a}{n}} = \frac{a}{n-a}.$$

Example 1. If there are 25 names in a hat, 15 girl's names and 10 boy's names, what are the odds favoring the random selection of a girl's name?

Solution:

$$\text{Odds favoring a girl's name} = \frac{P(G)}{P(\bar{G})} = \frac{\frac{15}{25}}{\frac{10}{25}} = \frac{15}{10} = \frac{3}{2}.$$

We say, then, that the odds favoring the selection of a girl's name are 3 to 2. (Similarly, the odds favoring the selection of a boy's name are 2 to 3.)

Example 2. If a penny is tossed, what are the odds in favor of its landing "heads up"?

Solution: Odds favoring $H = \dfrac{P(H)}{P(\overline{H})} = \dfrac{\frac{1}{2}}{\frac{1}{2}} = \dfrac{1}{1}$.

We say, then, that the odds favoring "heads up" are 1 to 1 or that the odds are even.

EXERCISE 8.6

1. If there are 3 red, 4 white, and 5 blue marbles in a hat and one is chosen at random, what is the probability of:
   (a) Choosing a red marble?
   (b) Not choosing a red marble?
   (c) Choosing a white or a blue marble?
   (d) Not choosing a red or a white marble?
   (e) Choosing a red, a white, or a blue marble?
   (f) Not choosing a red, a white, or a blue marble?

2. Assume a pair of dice is rolled:
   (a) In how many ways may the sum of the face numbers be 6?
   (b) What is the probability that the sum of the face numbers will be 6?

3. Assume a pair of dice is rolled:
   (a) What is the probability of rolling a 4 and a 2?
   (b) What is the probability of rolling two 3's?

4. If a committee of 5 is chosen at random from among 6 Democrats and 4 Republicans, what is the probability its members will be:
   (a) All Democrats?
   (b) All Republicans?
   (c) 3 Democrats and 2 Republicans?

5. If a pair of dice is rolled:
   (a) What is the probability that the sum of the face numbers will be 7?
   (b) What is the probability that the sum of the face numbers will not be 7?
   (c) What are the odds that the sum of the face numbers will be 7?
   (d) What are the odds that the sum of the face numbers will not be 7?

6. In the game of roulette described in the text (Example 5 of Section 8.5),

assume the casino wins when 0 or 00 is selected. Suppose a player bets that some even number from 1–36 will be selected:
(a) What is the probability of the player's winning?
(b) What is the probability of his not winning?
(c) What are the odds favoring the player?
(d) What are the odds favoring the casino?

7. If two cards are drawn without replacement from a standard 52-card deck (26 red cards and 26 black cards), what is the probability that:
(a) Both cards will be red?
(b) One card will be red and the other black?
(c) Both cards will be aces (there are 4 aces in a deck)?

8. There are 5 white marbles and 3 black marbles in a hat. If 4 marbles are selected at random, what is the probability of selecting:
(a) 2 white marbles and 2 black marbles?
(b) 3 white marbles and 1 black marble?

8.7 Mathematical Expectation

To introduce the very interesting topic of mathematical expectation let us look back at Example 5 of Section 8.5. This example concerned a particular wager that may be made in playing roulette. If a guest of the casino bets that an odd number will be selected, his probability of winning is $\frac{18}{38}$ or $\frac{9}{19}$. If the guest plays roulette only one time, he will obviously either win or lose; but if a guest plays roulette over a long period of time, what may he anticipate his average losses or winnings per game to be? The answer to this question as well as other similar questions is found by applying the concept of *mathematical expectation*.

Mathematical expectation, E, is the product of the probability that an event will occur and the numerical value of the occurrence. In the roulette example referred to above, if the guest of the casino bets $1.00 that an odd number will be selected, he will receive $2.00 if he wins (his own dollar plus one from the casino). His mathematical expectation is by definition:

$$E = \tfrac{9}{19} \times 2.00 \approx 0.95.$$

We see, then, that the guest has a mathematical expectation of about $.95. Therefore, if he paid 95 cents to play the game it would be a "fair" bet. However, since he pays $1.00 to play the game and his mathematical expectation is only $.95, he may anticipate losing an average of 5 cents for each dollar he bets over a period of time. By the same reasoning process, the casino may expect to win 5 cents for each dollar bet over a period of time or 5% of all bets. If, for example, an average of $10.00 is bet each minute on a particular roulette wheel, the bets for the day would total $60 \times 24 \times \$10 = \$14,400$. Hence, the casino could anticipate a net of about 5% of $14,400 or $720.

Example 1. Suppose 800 tickets are sold and the stubs from each are placed in a drum and thoroughly mixed. If a winning ticket is then chosen at random and its holder receives $600:

(a) What is the mathematical expectation of any given person who holds a ticket?
Solution: $E = \frac{1}{800} \times \$600 = \$0.75$.

(b) What price should be paid for a ticket if the game is "fair"?
Solution: Since the mathematical expectation is $0.75, the price of a ticket should be 75 cents.

(c) If $1.00 is paid for each ticket, what will be the average loss to the holder of each ticket?
Solution: $\$1.00 - \$0.75 = \$0.25$.

Sometimes an event may have more than one possible outcome. If the probability of each outcome is p_1, p_2, p_3, and so on, and the numerical value of the occurrence of each outcome is v_1, v_2, v_3, and so forth, then the mathematical expectation, E, is defined as:

$$E = p_1 v_1 + p_2 v_2 + p_3 v_3 + \cdots$$

Example 2. Suppose in a certain game you roll an ordinary die with the numbers 1, 2, 3, 4, 5, and 6 on its faces. If you are to receive in dollars an amount equal to the number on the upper face of the die:

(a) What is your mathematical expectation?
Solution:

$$\begin{aligned}
E &= p_1 v_1 + p_2 v_2 + p_3 v_3 + \cdots \\
&= \tfrac{1}{6}(1) + \tfrac{1}{6}(2) + \tfrac{1}{6}(3) + \tfrac{1}{6}(4) + \tfrac{1}{6}(5) + \tfrac{1}{6}(6) \\
&= \tfrac{1}{6} + \tfrac{2}{6} + \tfrac{3}{6} + \tfrac{4}{6} + \tfrac{5}{6} + \tfrac{6}{6} = \tfrac{21}{6} = 3.5
\end{aligned}$$

Hence, your mathematical expectation is $3.50.

(b) If you paid $4.00 for the privilege of playing this game, about what amount could you anticipate losing if you played the game 100 times?
Solution: The loss per game over a period of time could be expected to average $\$4.00 - \$3.50 = \$0.50$. Therefore, if the game is played 100 times, $100 \times \$0.50 = \$50.00$ would be a good guess of the total loss.

Example 3. Suppose a coin is tossed twice and you are to receive $5 if two heads are obtained, $3 if one head is obtained, and nothing if no heads are obtained. What is your mathematical expectation?

Solution: Since the sample space is $\{HH, HT, TH, TT\}$, we see that the probability of two heads is $\frac{1}{4}$, of one head is $\frac{1}{2}$, and of no heads is $\frac{1}{4}$. Hence, the mathematical expectation is $2.75 as shown by the following computations:

$$E = \tfrac{1}{4}(5) + \tfrac{1}{2}(3) + \tfrac{1}{4}(0) = \tfrac{5}{4} + \tfrac{3}{2} + 0 = \tfrac{11}{4} = 2.75.$$

In the introduction to this chapter, we mentioned that Pascal was one of the early contributors to the theory of probability. After Pascal's interest turned from mathematics to religion, he actually used mathematical expectation to help support his decision to lead a religious life. His argument was that even though the probability that a religious life will lead to heaven may be very small, the eternal happiness of heaven is an infinite winning. Therefore, the mathematical expectation, which is a product of these two factors, is also infinite and well worth the risk.

EXERCISE 8.7

1. A man is to receive $5.00 if he can roll a pair of dice once and have the sum of the face numbers total 7.
   (a) What is the probability he will succeed?
   (b) What is his mathematical expectation?
   (c) If he pays $1.00 for the privilege of rolling the dice, how much may he expect his average loss per game to be over a period of time?

2. (a) A die is rolled. If you select in advance the face that lands up, then you win $12. What is your mathematical expectation?
   (b) If you pay $2.50 each time the game is played, how much may you expect to lose if you play 100 games?

3. There are 4 boxes. One of the boxes contains 20-dollar bills, another 10-dollar bills, another 5-dollar bills, and another 1-dollar bills. If you may select a box at random and keep a bill that you draw from it, what would be a fair price for the privilege of playing this game?

4. (a) If a single die is rolled, what is the expected number of spots on the upper face?
   (b) If a die is rolled 20 times, what is the expected sum of the spots that will land face up?
   (c) If you have a die, roll it 20 times and see how close your result is to the expected result obtained in part (b).

5. In a lottery, one thousand tickets are sold at $2.00 each. Four tickets are chosen at random in determining a first prize of $400, a second of

$300, a third of $200, and a fourth prize of $100. What is the mathematical expectation of a person holding:
(a) One ticket?
(b) Two tickets?
(c) One hundred tickets?
(d) None of the tickets?
(e) All of the tickets?

8.8 Coin Tossing and Pascal's Triangle

Let us consider once again the notion of coin tossing. Our object will be to determine the total number of possible ways of having 0, 1, 2, 3, . . . tails face up according to the number of coins tossed. The possible results of tossing one coin n times, of course, are the same as tossing n coins one time. First let us toss 1 coin. The possible results are obviously 0 tails or 1 tail as shown below:

H	*T*
1 case	1 case

If 2 coins are tossed (or one coin is tossed twice) and the possible results are grouped according to 0, 1, or 2 tails face up, we have

HH	*HT TH*	*TT*
1 case	2 cases	1 case

Tossing 3 coins we have the possibility of 0, 1, 2, or 3 tails face up. The results follow:

HHH	*HHT HTH THH*	*HTT THT TTH*	*TTT*
1 case	3 cases	3 cases	1 case

Tossing 4 coins we have according to the number of possible tails face up:

	HHHT HHTH	*HHTT HTHT HTTH*	*HTTT THTT*	
HHHH	*HTHH THHH*	*THHT THTH TTHH*	*TTHT TTTH*	*TTTT*
1 case	4 cases	6 cases	4 cases	1 case

A continuation of this process to consider all possible results of tossing various numbers of coins would be tedious and boring. However, let us summarize the results from above and look for a possible relationship between the number of coins tossed and the number of ways 0, 1, 2, 3, . . . tails may be face up.

Number of Coins	Number of Ways of Obtaining:				
	0 Tails	1 Tail	2 Tails	3 Tails	4 Tails
1	1	1			
2	1	2	1		
3	1	3	3	1	
4	1	4	6	4	1

By rearranging the numbers in the above table in an array as follows, we obtain what is commonly called **Pascal's triangle**.

Number of Coins	Number of Ways of Obtaining 0, 1, 2, 3, . . . Tails					
1			1	1		
2		1	2	1		
3	1	3	3	1		
4	1	4	6	4	1	

Note that each row begins and ends with 1, and that any interior number may be found by adding the two numbers diagonally above it in the previous row. Following this technique we can extend Pascal's triangle indefinitely; below we annex two more rows.

Number of Coins	Number of Ways of Obtaining 0, 1, 2, 3, . . . Tails						
1				1	1		
2			1	2	1		
3		1	3	3	1		
4	1	4	6	4	1		
5	1	5	10	10	5	1	
6	1	6	15	20	15	6	1

If a coin is tossed, it may land in exactly 2 ways, head up or tail up. Hence, by the fundamental counting principle, if n coins are tossed they may land in 2^n different ways. Examining Pascal's triangle, we can verify this statement by adding the numbers in each row as shown below.

Number of Coins	Sum of the Number of Ways of Obtaining 0, 1, 2, 3, . . . Tails
1	$1 + 1 = 2 = 2^1$
2	$1 + 2 + 1 = 4 = 2^2$
3	$1 + 3 + 3 + 1 = 8 = 2^3$
4	$1 + 4 + 6 + 4 + 1 = 16 = 2^4$
5	$1 + 5 + 10 + 10 + 5 + 1 = 32 = 2^5$
6	$1 + 6 + 15 + 20 + 15 + 6 + 1 = 64 = 2^6$

Example 1. If 6 coins are tossed:

(a) In how many ways can 4 tails be obtained?

Solution: $C(6, 4) = \dfrac{P(6, 4)}{4!} = \dfrac{6 \cdot 5 \cdot 4 \cdot 3}{4 \cdot 3 \cdot 2 \cdot 1} = 15.$

(b) In how many ways can 2 heads be obtained?

Solution: $C(6, 2) = \dfrac{P(6, 2)}{2!} = \dfrac{6 \cdot 5}{2 \cdot 1} = 15.$

Note in Example 1 that the number of ways of obtaining 4 tails is the same as the number of ways of obtaining 2 heads. This is to be expected since, if 6 coins are tossed and 4 land tails up, then there must be $6 - 4$ or 2 that land heads up. Following this reasoning, we would expect $C(6, 0) = C(6, 6)$; $C(6, 1) = C(6, 5)$; $C(6, 2) = C(6, 4)$; and so on. This is indeed the case, and for this reason each row of Pascal's triangle is symmetric about its center. Pursuing this line of reasoning even further, it can be proved that, in general, $C(n, r) = C(n, n - r)$.

Example 2. Show that $C(10, 7) = C(10, 10 - 7)$ or $C(10, 3)$.

Solution:

$$C(10, 7) = \frac{P(10, 7)}{7!} = \frac{10 \cdot 9 \cdot 8 \cdot 7 \cdot 6 \cdot 5 \cdot 4}{7 \cdot 6 \cdot 5 \cdot 4 \cdot 3 \cdot 2 \cdot 1} = 120.$$

$$C(10, 3) = \frac{P(10, 3)}{3!} = \frac{10 \cdot 9 \cdot 8}{3 \cdot 2 \cdot 1} = 120.$$

Suppose we wish to use Pascal's triangle directly in computing, for example, the possible number of ways of obtaining 0 through 70 tails face up when 70 coins are tossed; then it is necessary to fill in the 69 rows that precede the relevant one. Is it possible to find one row of Pascal's triangle without first finding those that precede it? The answer is yes, as we will illustrate by considering the tossing of 4 coins. If 4 coins are tossed, the possible number with tails face up is 0, 1, 2, 3, or 4. We are interested in the number of ways in which each of these outcomes may occur. Since we wish to know the number of different sets of coins that may be selected to be tails face up, and since the order in which we choose these coins is irrelevant, we are dealing with combinations rather than permutations. Specifically, in considering the number of ways in which 0, 1, 2, 3, or 4 tails may be face up in the tossing of 4 coins, we need to find $C(4, 0)$, $C(4, 1)$, $C(4, 2)$, $C(4, 3)$, and $C(4, 4)$. Since there is only one way in which 0 tails may occur (all heads up), it must be true that $C(4, 0) = 1$. This is indeed the case, for, in general

$$C(n, 0) = \frac{P(n, 0)}{0!} = \frac{1}{1} = 1.$$

(Note that our previous definitions of $0! = 1$ and $P(n, 0) = 1$ give us the result we desire for $C(n, 0)$.) Following are all of the computations necessary for determining the number of ways in which 0, 1, 2, 3, or 4 tails may be face up when 4 coins are tossed:

$$C(4, 0) = \frac{P(4, 0)}{0!} = \frac{1}{1} = 1$$

$$C(4, 1) = \frac{P(4, 1)}{1!} = \frac{4}{1} = 4$$

$$C(4, 2) = \frac{P(4, 2)}{2!} = \frac{4 \cdot 3}{2 \cdot 1} = 6$$

$$C(4, 3) = \frac{P(4, 3)}{3!} = \frac{4 \cdot 3 \cdot 2}{3 \cdot 2 \cdot 1} = 4$$

$$C(4, 4) = \frac{P(4, 4)}{4!} = \frac{4 \cdot 3 \cdot 2 \cdot 1}{4 \cdot 3 \cdot 2 \cdot 1} = 1$$

The final results from above are 1, 4, 6, 4, and 1, which constitute the elements of one of the rows in Pascal's triangle. Therefore, Pascal's triangle can be rewritten entirely in terms of combinations and it is a simple matter to determine an arbitrary row. The first few rows are shown below—which one contains the elements 1, 4, 6, 4, and 1?

$$C(1, 0) \qquad C(1, 1)$$
$$C(2, 0) \qquad C(2, 1) \qquad C(2, 2)$$
$$C(3, 0) \qquad C(3, 1) \qquad C(3, 2) \qquad C(3, 3)$$
$$C(4, 0) \qquad C(4, 1) \qquad C(4, 2) \qquad C(4, 3) \qquad C(4, 4)$$
$$C(5, 0) \qquad C(5, 1) \qquad C(5, 2) \qquad C(5, 3) \qquad C(5, 4) \qquad C(5, 5)$$

Example 3.   Assume a coin is tossed 5 times:
(a) In how many different ways may the coin land?
Solution: $2^5 = 32$.

(b) In how many ways may 3 heads be obtained?

Solution: $C(5, 3) = \dfrac{P(5, 3)}{3!} = \dfrac{5 \cdot 4 \cdot 3}{3 \cdot 2 \cdot 1} = 10.$

(c) What is the probability of obtaining 3 heads?

Solution: $P(3 \text{ heads}) = \dfrac{C(5, 3)}{2^5} = \dfrac{10}{32} = \dfrac{5}{16}.$

EXERCISE 8.8

1. Assume 7 coins are tossed:
   (a) In how many ways may 5 heads be obtained?
   (b) In how many ways may 2 tails be obtained?

2. Show that $C(7, 3) = C(7, 7 - 3)$ or $C(7, 4)$.

3. Find the fifth row of Pascal's triangle by computing $C(5, 0)$, $C(5, 1)$, $C(5, 2)$, $C(5, 3)$, $C(5, 4)$, and $C(5, 5)$.

4. (a) If 8 coins are tossed, in how many different ways may they land?
   (b) If one coin is tossed 8 times, in how many different ways may it land?

5. If 6 coins are tossed:
   (a) In how many different ways may they land?
   (b) In how many different ways may they land with 3 heads up?
   (c) What is the probability of obtaining 3 heads up?

6. If a coin is tossed 7 times:
   (a) In how many different ways may it land?
   (b) In how many ways may it land with 5 tails up?
   (c) What is the probability of obtaining 5 tails up?

7. The sixth row of Pascal's triangle has elements 1, 6, 15, 20, 15, 6, and 1. Using the elements of the sixth row, write the elements of the next two rows.

8. Write the fifth, sixth, and seventh rows of Pascal's triangle in terms of combinations.

8.9 The Binomial Theorem (Optional Section)

A mathematical expression with two terms or addends such as $3x^2 + 4y$ or $2x^3y + 7x^2y^6$ is called a **binomial**. Binomials and successive factors of binomials are used extensively in various branches of mathematics. If we use the expression $(a + b)$ to indicate a binomial, then $(a + b)^2$ would indicate that we have two factors of the binomial $(a + b)$. If the distributive property is applied, we may write $(a + b)^2$ in **expanded form** as follows:

$$\begin{aligned}
(a + b)^2 &= (a + b)(a + b) \\
&= (a + b)(a) + (a + b)(b) \\
&= a^2 + ba + ab + b^2 \\
&= a^2 + 2ab + b^2
\end{aligned}$$

By using the distributive property to expand $(a + b)^2$, the effect is equivalent to selecting an a or b in all possible ways from each binomial factor, multiplying the selected terms, and finally adding the results thus obtained. Employing this procedure we have

$$(a + b)^2 = (a + b)(a + b) = a^2 + ab + ba + b^2 = a^2 + 2ab + b^2.$$

Analogously, we find that

$$\begin{aligned} (a + b)^3 &= (a + b)(a + b)(a + b) \\ &= (a^2 + 2ab + b^2)(a + b) \\ &= a^3 + a^2b + 2a^2b + 2ab^2 + b^2a + b^3 \\ &= a^3 + 3a^2b + 3ab^2 + b^3 \end{aligned}$$

Note that in the expanded form of $(a + b)^3$ the number of factors of a's and b's in each term totals 3. Similarly, the number of a's and b's in each term of the expansion of $(a + b)^4$ totals 4 as can be seen in the equation below:

$$(a + b)^4 = a^4 + 4a^3b + 6a^2b^2 + 4ab^3 + b^4.$$

In general, each term of the expanded form of $(a + b)^n$ will have a total of n factors of a and b. This is to be expected, of course, since each addend for the expansion may be found by selecting an a or b from each of the n factors of $(a + b)$ and multiplying them together.

Given an expression such as $a^4 + 4a^3b + 6a^2b^2 + 4ab^3 + b^4$, we call the numerical factors 1, 4, 6, 4, and 1 the **coefficients** of a^4, a^3b, a^2b^2, ab^3, and b^4 respectively. Let us examine the coefficients in the following display. (Recall that by definition $x^0 = 1$, provided $x \neq 0$.)

$$\begin{aligned} (a + b)^0 &= 1 \\ (a + b)^1 &= a + b \\ (a + b)^2 &= a^2 + 2ab + b^2 \\ (a + b)^3 &= a^3 + 3a^2b + 3ab^2 + b^3 \\ (a + b)^4 &= a^4 + 4a^3b + 6a^2b^2 + 4ab^3 + b^4 \\ (a + b)^5 &= a^5 + 5a^4b + 10a^3b^2 + 10a^2b^3 + 5ab^4 + b^5 \end{aligned}$$

The coefficients of the various terms of the expansions above will be found to constitute Pascal's triangle. In fact, since $(a + b)^0 = 1$, the top vertex of the triangle, which was previously omitted, is now included. Writing only the coefficients of the terms we have the triangle that is shown on the top of the following page.

$(a + b)^0$						1						
$(a + b)^1$						1		1				
$(a + b)^2$					1		2		1			
$(a + b)^3$				1		3		3		1		
$(a + b)^4$			1		4		6		4		1	
$(a + b)^5$		1		5		10		10		5		1
$(a + b)^6$	1		6		15		20		15		6	1

Probably one would not expect the tossing of coins and the expansion of a binomial expression to be related. Upon reflection, however, a relationship becomes obvious. In tossing 6 coins, as we observed previously, the total number of ways of obtaining 4 tails is $C(6, 4) = 15$. Of course, if there are 4 tails, there must be 2 heads since $4 + 2 = 6$. Hence the number of ways of obtaining 4 tails by tossing 6 coins is the same as the number of ways of obtaining 2 heads, which is $C(6, 2)$; and we see that $C(6, 4) = C(6, 2) = 15$. In the expansion of $(a + b)^6$, what will be the coefficient of a^4b^2? The expansion of $(a + b)^6$ involves selecting either an a or a b from each of the 6 factors of $(a + b)^6$ and indicating their product. The coefficient of the a^4b^2 term will correspond to the number of ways in which 4 a's and 2 b's can be selected from among the 6 factors of $(a + b)^6$, which is $C(6, 4)$. Selecting 4 a's for the a^4b^2 term, however, is equivalent to selecting 2 b's, which may be done in $C(6, 2)$ different ways. Consequently, the coefficient of the term containing the expression a^4b^2 is $C(6, 4) = C(6, 2) = 15$. Finally, we see that the number of ways of obtaining 4 tails and 2 heads when 6 coins are tossed is the same as the number of terms having 4 a's and 2 b's in the expansion of $(a + b)^6$. This analogy, of course, may be extended to any number of coins n and any term of the expansion of $(a + b)^n$.

Note that in general, the expansion of $(a + b)^n$ may be written as follows:

$$(a + b)^n = C(n, 0)a^n + C(n, 1)a^{n-1}b + C(n, 2)a^{n-2}b^2 + \cdots + C(n, n)b^n.$$

This equation is appropriately called the **binomial theorem**. Note that the coefficients in the binomial theorem have been chosen to represent the number of ways of choosing the factors of b in each term rather than the ways of selecting the factors of a. This decision, of course, is purely arbitrary since the number of ways of selecting the b's for any given term is the same as the number of ways of selecting the a's; for, in general, $C(n, r) = C(n, n - r)$.

Example 1. Expand $(a + b)^4$ by using the binomial theorem.

Solution:

$$(a + b)^4 = C(4, 0)a^4 + C(4, 1)a^3b + C(4, 2)a^2b^2 + C(4, 3)ab^3 + C(4, 4)b^4$$

$$= \frac{1}{0!}a^4 + \frac{4}{1!}a^3b + \frac{4 \cdot 3}{2!}a^2b^2 + \frac{4 \cdot 3 \cdot 2}{3!}ab^3 + \frac{4 \cdot 3 \cdot 2}{4!}b^4$$

$$= a^4 + 4a^3b + 6a^2b^2 + 4ab^3 + b^4$$

Example 2. Show that $5^4 = (2 + 3)^4$ by using the binomial theorem.

Solution: $5^4 = 5 \cdot 5 \cdot 5 \cdot 5 = 625$.

$$(2 + 3)^4 = C(4, 0)2^4 + C(4, 1)2^3 \cdot 3 + C(4, 2)2^2 \cdot 3^2 + C(4, 3)2 \cdot 3^3$$
$$+ C(4, 4)3^4$$
$$= (1)2^4 + 4(2^3 \cdot 3) + 6(2^2 \cdot 3^2) + 4(2 \cdot 3^3) + 1(3^4)$$
$$= 16 + 4(24) + 6(36) + 4(54) + 81$$
$$= 16 + 96 + 216 + 216 + 81$$
$$= 625$$

Example 3. Expand $(2x^2 + 3y)^3$ by applying the binomial theorem.

Solution:

$$(2x^2 + 3y)^3 = C(3, 0)(2x^2)^3 + C(3, 1)(2x^2)^2(3y)$$
$$+ C(3, 2)(2x^2)(3y)^2 + C(3, 3)(3y)^3$$
$$= 1(8x^6) + 3(4x^4)(3y) + 3(2x^2)(9y^2) + 1(27y^3)$$
$$= 8x^6 + 36x^4y + 54x^2y^2 + 27y^3$$

EXERCISE 8.9

1. Use the binomial theorem to expand:
   (a) $(a + b)^4$ (b) $(x + y)^5$ (c) $(5x + y^2)^3$ (d) $(xy + 4)^3$

2. In the expansion of $(a + b)^8$:
   (a) How many terms are there?
   (b) Which term contains a^5b^3?
   (c) What is the coefficient of the term containing a^5b^3?

3. In the expansion of $(a + b)^6$:
   (a) What is the coefficient of the term containing a^4b^2?
   (b) Which other term has the same coefficient as the term containing a^4b^2?

4. Show that $2^4 = (1 + 1)^4$ by using the binomial theorem.

5. (a) Approximate $(1.02)^{10}$ by finding the sum of the first 4 terms of $(1 + .02)^{10}$. (Note that each term is smaller than the preceding one.)
   (b) The approximation of $(1.02)^{10}$ accurate to 6 digits is 1.21899. How much larger is this than your approximation in part (a)?

(c) How could you get a more accurate approximation of $(1.02)^{10}$?

(d) How could the exact value of $(1.02)^{10}$ be obtained?

REVIEW EXERCISES

1. (a) If 5 different house models are available in a development and each can be purchased in any of 7 different color schemes, how many different purchases are possible?

   (b) A test has 10 statements, each of which may be marked true, marked false, or omitted altogether. In how many ways may such a test be answered?

2. (a) In how many different ways may 5 people be seated in 5 chairs?

   (b) In how many different ways may 3 people be seated in 5 chairs?

3. (a) In how many ways may a committee with 3 members be selected from among 7 people?

   (b) In how many ways may a committee of 3 members be selected from among 7 people if one member is to be designated as chairman?

4. In how many ways may a president, vice-president, and treasurer be selected from among 12 members of a club?

5. If the letters must come first, how many different 7-symbol license tags may be made by rearranging the letters A, B, C and the numerals 3, 4, 5, 6?

6. If a committee of 3 is chosen at random from among 6 girls and 4 boys, what is the probability the committee will:

   (a) Contain all boys?

   (b) Contain 2 girls and 1 boy?

7. If a pair of dice is rolled, what is the probability that:

   (a) The sum of the faces will be 6?

   (b) The sum of the faces will be less than 7?

   (c) The sum of the faces will be an odd number?

   (d) The sum of the faces will be an even number?

8. Assume a pair of dice is rolled; what are the odds that:

   (a) The sum of the faces will be 5?

   (b) The sum of the faces will not be 5?

   (c) The sum of the faces will be less than 7?

   (d) The sum of the faces will be greater than or equal to 7?

9. In a certain game you may roll a pair of dice and receive $30 if the sum of the faces totals 7. What would be a "fair" price to pay for playing such a game? Explain.

10. If a coin is tossed 6 times:

    (a) In how many different ways may the coin land?

    (b) In how many ways may 4 heads be obtained?

    (c) What is the probability of obtaining 4 heads?

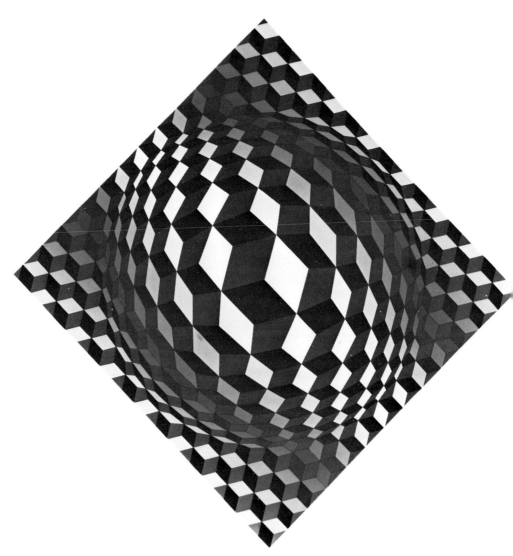

Victor Vasarely, *Cheyt M*
1970. Tempera on canvas. 107 x 106¼" (lozenge).
The Solomon R. Guggenheim Museum, New York

⬤ **TOPOLOGY**

⬤.⬤ Introduction

A geometric figure such as a line segment, circle, or triangle may be thought of as being rigid, and we may imagine moving it about in space without changing its shape or distorting it in any way. If a triangle, for example, is considered to be rigid and we move it in space, the lengths of the sides will remain the same, the measures of the angles will remain the same, the sides will not bend, the area will be constant, and the distance between any two specified points will not vary. Properties that do not change, such as shape, length, area, and so on, are said to be *invariant*.

Suppose, however, a triangle is considered to be composed of rubber which is "completely elastic" and can be stretched or contracted as much as we desire. If this triangle is moved about in space, the sides are bent and stretched, the angles are distorted, and so on, are there any properties that still are invariant? The answer is "yes." For example, regardless of how the elastic triangle is stretched or bent, it must be cut twice if it is to fall into two distinct pieces. The study of such invariant properties under *elastic motion* belongs to that branch of mathematics called topology (from the Greek *topo*, space; and *logia*, study).

In topology, any figure may be subjected to elastic motion, which

207

consists of bending, stretching, shrinking, and so on. In general, elastic motion permits distorting a figure in any way whatsoever, provided that distinct points remain distinct; that is, no two separate points are allowed to merge into a single point. For example, elastic motion does not allow us to change a circle into a figure 8; for, in making the transformation, two distinct points of the circle must merge into a single point at the center of the 8. Similarly, a figure 8 may not be transformed into a circle by cutting it at its central point. Elastic motion includes cutting a figure only if we "sew up" the cut exactly as it was before. For example, we may cut across a circular belt, give one end a full twist (360°) and sew up the cut again. However, elastic motion does not permit giving the belt a half-twist (180°) and sewing it up since points along the cut edges would not be matched together as they were prior to cutting. If, through elastic motion, one figure can be made to coincide with another, the two figures are said to be **topologically equivalent**.

We will begin our study by considering figures of a plane that are topologically equivalent. Imagine the surface of a desk top as a plane and a rubber band of no thickness lying in the plane. All of the various shapes into which we may distort the rubber band through elastic motion without its leaving the plane represent a set of topologically equivalent plane figures. There are, of course, infinitely many such figures in this set, twelve of which are represented in Figure 9.1.1.

Each of the twelve figures of Figure 9.1.1 is a **simple closed curve**. A simple closed curve has the topological property of partitioning the plane to which it belongs into three disjoint subsets: The set of points of the curve itself, the set of points called the **interior** of the curve, and the set of points called the **exterior** of the curve. In Figure 9.1.2, the interior of a simple

FIGURE 9.1.1

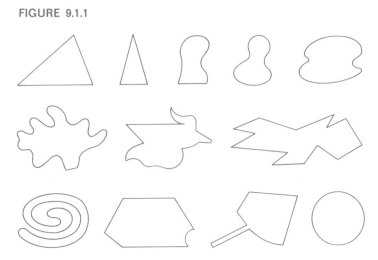

FIGURE 9.1.2

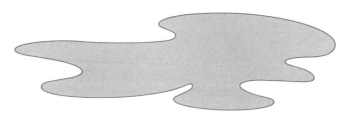

closed curve has been shaded. Note that a *simple* closed curve has only *one* interior. All simple closed curves are, of course, topologically equivalent. Other examples of topologically equivalent geometric figures are shown in Figure 9.1.3; the figures of part (a) are equivalent, those of part (b) are equivalent, and so on. Observe that either member of a pair of equivalent figures may be transformed into the other by elastic motion.

FIGURE 9.1.3

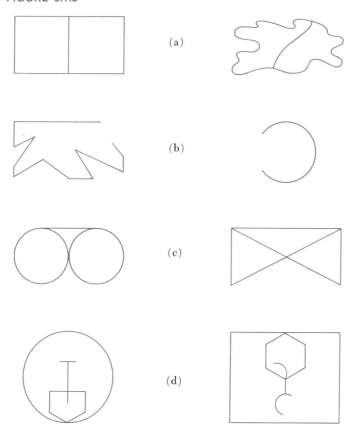

Until now we have focused our attention primarily on figures in a plane; these figures are two-dimensional and have both length and width. If we consider three-dimensional figures, which may have height as well as length and width, there are many more topologically equivalent figures that can be studied. For example, the open box, the punctured sphere, the conical drinking cup, and the circular disc in Figure 9.1.4 are topologically equivalent since each may be obtained from any one of the others by elastic motion. The fact that the circular disc is a plane (two-dimensional) figure does not prohibit it from being topologically equivalent to the other figures that are three-dimensional. With perhaps considerable stretching, pushing, and squeezing, you can imagine any of the given three-dimensional figures being flattened and distorted into the shape of a circular disc.

FIGURE 9.1.4

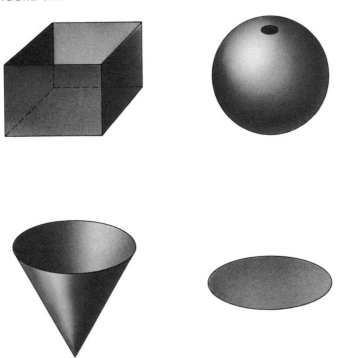

Analogous to a simple closed figure in two dimensions is the **simple closed surface** in three dimensions. An example of a simple closed surface is the sphere, as represented by a balloon or a tennis ball. The simple closed surface partitions space into three disjoint subsets: The set of points of the surface itself, the set of points inside or *interior* to the surface, and the set of points outside or *exterior* to the surface. A *simple* closed surface has

FIGURE 9.1.5

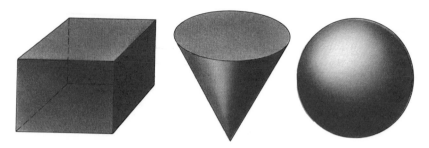

only *one* interior. Some simple closed surfaces are the closed box or cube, the conical drinking cup with a lid on it, and the sphere (see Figure 9.1.5).

Each of the figures in Figure 9.1.5 is topologically equivalent to the others, but none of them is equivalent to those of Figure 9.1.4. To show that this is true, suppose we transform the punctured sphere into an unpunctured sphere. In closing the puncture some of the points along the perimeter of the puncture would have to merge with other points along the perimeter. In fact, we could make the puncture smaller and smaller until all points on the perimeter came together at a single point. Elastic motion, however, does not permit distinct points to come together. The rules of elastic motion are also violated by cutting (and not resewing) the unpunctured sphere to make it coincide with the punctured sphere.

Our comments so far would seem to indicate that topology is essentially a branch of geometry, and to some extent this is true. Topology was first studied as a branch of geometry, but during the last 40 or 50 years it has become so generalized and involved with other branches of mathematics that it is probably best to give it a separate classification.

Number theory and topology are two branches of mathematics where even the amateur can have a lot of fun—yet number theory is one of the oldest areas of mathematics while topology is one of the newest. Most of the work in topology has been done within the last century. However, a few interesting and important concepts were developed earlier.

Among the first persons to develop concepts in the field of topology were René Descartes (1596–1650), Leonard Euler (1707–1783), Carl F. Gauss (1777–1855), and August F. Möbius (1790–1868). Bernard Riemann (1826–1866) and Henri Poincaré (1854–1912) also rank high among relatively early contributors to topology. More recent discoveries have come from such men as Maurice Fréchet, Oswald Veblen, J. W. Alexander, L. E. J. Brower, and Solomon Lefschetz. Within the last hundred years there have been so many contributors to the field of topology that it almost seems unjust to mention some without naming others. Important topological concepts have been developed by mathematicians in Italy,

Poland, Russia, France, Switzerland, Germany, and the United States to name a few representative countries. Today a sizable group of mathematicians throughout the world are devoting their efforts to topology.

EXERCISE 9.1

1. The following six figures contain three pairs that are topologically equivalent. Identify the equivalent pairs.

(a) (b) (c)

(d)  (e)  (f)

2. Find the topologically equivalent pairs:

(a) (b) (c)

(d) (e) (f)

3. Identify the topologically equivalent pairs in the following:

(a) (b)  (c)

(d) (e)  (f)

4. In the following diagram involving a simple closed curve, point *P* is known to be exterior to the curve. Each of the points *A, B, C, D, E,* and *F* are joined by a line segment with point *P*.
   (a) How many times does each of the line segments cross the simple closed curve?
   (b) Does there seem to be a relationship between the number of times the line segment crosses the curve and whether or not the point from which it originates is interior or exterior to the curve?

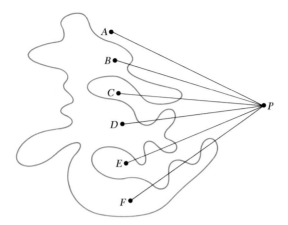

(c) Write the following sentence so it reads correctly: If an arbitrary point Q is joined by a line segment to a point P outside a simple closed curve, then Q is outside the curve if the line segment crosses the curve an (*odd, even*) number of times and inside the curve if the line segment crosses the curve an (*odd, even*) number of times.

5. Imagine that an ordinary ruler marked for measuring from 1 to 12 inches is made of rubber. If we stretch the ruler uniformly by pulling on both ends until the ruler is twice as long as it was originally:
   (a) How long will the ruler be?
   (b) What number on the ruler could be used in measuring a true distance of 6 inches? 12 inches?
   (c) What number on the ruler could be used in measuring a distance of 18 inches? 24 inches?
   (d) Will all of the numbers on the ruler be in the same order as before the ruler was stretched?

6. (a) Draw any simple closed curve on a sheet of paper and cut along the curve with a pair of scissors. Into how many distinct pieces is the paper divided?
   (b) On a sheet of paper draw any closed curve that crosses itself once (such as a figure 8). Cut along the curve with a pair of scissors. Into how many pieces is the paper divided?
   (c) Into how many pieces will a sheet of paper be divided if we cut along a closed curve that crosses itself 2 times? 3 times? n times?

7. (a) If a left-handed glove is turned inside out will it still fit the left hand?
   (b) Is a left-handed glove topologically equivalent to a right-handed glove?

8. (a) A surface the shape of a doughnut is called a *torus*. Does a torus have only one interior?
   (b) Can a torus be distorted into a sphere by elastic motion?
   (c) Is a torus topologically equivalent to a sphere?
   (d) What is meant when we say that the absent-minded topologist couldn't tell his doughnut from his coffee cup?

9. A circle is drawn on a "completely flexible" rubber sheet, and some point A inside the circle is joined by a line segment to some point B outside the circle. Can the sheet be distorted by elastic motion in such a way that:
   (a) The circle becomes an ellipse?
   (b) The circle becomes a square?
   (c) Point A moves to within a hundredth of an inch from point B?
   (d) Point A is 1,000,000 miles from point B?
   (e) Point A coincides with point B?
   (f) Point A is outside the circle?
   (g) Point B is inside the circle?
   (h) Point A is no longer joined to point B by the line segment?

9.2 Networks

Although many topics in topology are very generalized and frequently require lengthy, detailed, rigorous proofs, we will avoid formal proofs and rely a great deal on the student's geometric intuition to arrive at conclusions. Our attention will be focused primarily on a few topics that reflect topology's geometric origin. One such topic is the study of *networks*.

A very famous problem that stimulated interest in networks was the Königsberg bridge problem. In fact, it would probably not be a great exaggeration to say that the solution of this problem marked the beginning of mathematicians' interest in questions of a topological nature. In the eighteenth century the town of Königsberg was part of East Prussia. (It has since been renamed Kaliningrad and is presently part of Russia.) The town is uniquely situated along the branches of the Pregel River on four separate land masses, one of which is an island. In the eighteenth century the four separate parts of the city were connected by seven bridges as shown in Figure 9.2.1. The four parts of the city are labeled A, B, C, and D in the diagram. The people of Königsberg wondered whether or not it was possible to walk across each of the seven bridges without recrossing any of them.

The Königsberg bridge problem was solved in 1735 by the famous Swiss mathematician Leonard Euler. His solution, which was general enough to solve all similar problems, involved making a network diagram like the one in Figure 9.2.2. In Figure 9.2.2, the points A, B, C, and D correspond to the four land masses of Figure 9.2.1, and the paths joining them correspond to the seven bridges. If it is possible to draw the network of Figure 9.2.2 without lifting your pencil from the paper and without retracing any of the paths, then, of course, it is possible to walk across each of the seven bridges of Königsberg without recrossing any of them. You may like to try drawing a diagram similar to Figure 9.2.2 without lifting

FIGURE 9.2.1

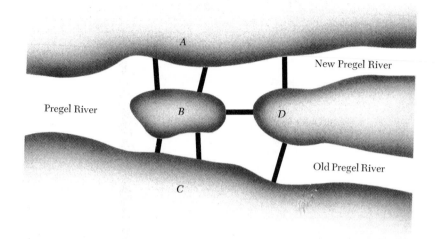

A

New Pregel River

Pregel River

B

D

Old Pregel River

C

your pencil from the paper and without retracing any of the paths. We will not state Euler's conclusion concerning this problem since you will be able to formulate your own answer after studying a few concepts concerning networks.

In the definition of a network we will use the term *edge*. An edge may be thought of as a finite path through space that does not cross itself. An edge has two end points that may or may not be distinct. In Figure 9.2.3, diagram (a) is not an edge because the path crosses itself, diagram

FIGURE 9.2.2

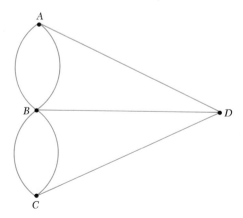

A

B

D

C

FIGURE 9.2.3

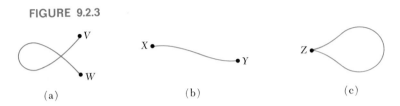

(a)   (b)   (c)

(b) is an edge with distinct end points X and Y, and diagram (c) is an edge with its two end points coinciding at Z.

A **network** consists of a nonempty finite set of edges; no two edges intersect except possibly at their end points. The end points of the edges in a network are called **vertices** (singular: vertex) and may be indicated by large dots. Each end point of an edge is a vertex in a network, but when two or more end points coincide they form a single vertex. Hence the number of distinct vertices in a network may be less than the total number of end points of the edges. Figure 9.2.4, for example, is a network of five edges (each of which has two end points), but there are only four vertices—A, B, C, and D. The network of Figure 9.2.4 is a **connected network** since it is possible to travel from any vertex to any other vertex by moving along its edges. In our discussions, we will consider only connected networks.

FIGURE 9.2.4

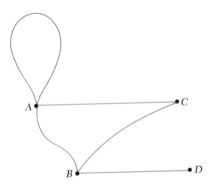

If it is possible to draw a network without lifting the pencil from the paper and without retracing any edge, then the network is **traversable**. (The solution to the Königsberg bridge problem consists of determining whether or not Figure 9.2.2 is traversable.) In determining the conditions under which a network is traversable, we will find it convenient to classify each vertex as *odd* or *even* according to whether it has an odd or even number of **edge emanations**. Each edge is said to **emanate** from the vertices (or vertex) where its end points are located. Since each edge has two end points, it will either emanate from each of two distinct vertices or it will emanate *twice* from the same vertex in the event that the edge forms a loop. For example, in Figure 9.2.4 the number of edge emanations from vertices A, B, C, and D respectively are 4, 3, 2, and 1. Hence vertices A and C are even vertices, while B and D are odd vertices. The total number of edge emanations in a network of n edges is $2n$, an even number. Thus the network of Figure 9.2.4 has 5 edges and $2 \times 5 = 10$ edge emanations from its vertices.

Since the total number of edge emanations of a network must be an even number, the number of odd vertices must be even. (Recall that the sum of two odd numbers is an even number, but that the sum of an odd and an even number is odd.) Each vertex in the networks of Figure 9.2.5 is labeled according to its number of edge emanations. Observe that in each network the sum of the edge emanations is twice the sum of the edges and that there are an even number of odd vertices.

FIGURE 9.2.5

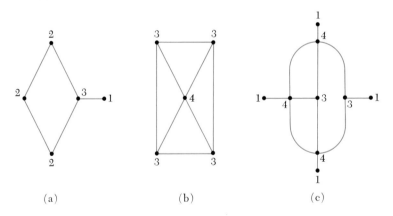

(a) (b) (c)

Suppose we are attempting to traverse a network that has only even vertices by beginning at some arbitrary vertex A as illustrated in Figure 9.2.6. If we traverse all edges emanating from A, they will be traversed in pairs by leaving A on one edge and eventually returning on another. Since there are a finite number of pairs of edges emanating from A, we will be unable to leave A after returning to it on its last untraversed edge. Hence if any network with only even vertices is indeed traversable and our trip begins at a given vertex A, it must end at A; it is important, therefore, not to return to A on the only untraversed edge leading to it until all other edges of the network have been traversed. If our trip begins

FIGURE 9.2.6

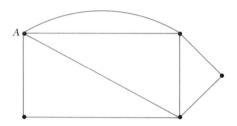

and ends at some arbitrary vertex A, this means that all edges emanating from other even vertices may be traversed in pairs by arriving on one edge of a pair and escaping on the other. Therefore, we will never be trapped at a vertex and unable to return to the beginning vertex of our trip. A little more reasoning leads to the conclusion that *any network that has only even vertices may be traversed by starting and ending at any given vertex*. Care must be taken, however, not to isolate some untraversed portion of the network by traversing all edges that lead to it and returning to the starting point too soon. The reader will find that the network of Figure 9.2.6 has all even vertices and can, therefore, be traversed by starting and ending at any given vertex.

Since the number of odd vertices in a network is always even, let us consider whether or not a network with exactly two odd vertices is traversable. Figure 9.2.7 will be used as an example. As previously explained, if we begin our trip at an even vertex such as C of Figure 9.2.7, our trip must end at that vertex. However, we will be unable to conclude our trip at C since eventually we will either be trapped at one of the odd

FIGURE 9.2.7

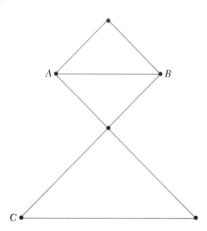

vertices or fail to traverse all of the edges leading to it. The reason for becoming trapped at one of the odd vertices becomes apparent when it is realized that in traversing the edges emanating from an odd vertex, they will be traversed in pairs by arriving at the vertex on one edge and leaving on another; eventually there will be only one edge leading to the odd vertex and we will be unable to leave after traversing it. In Figure 9.2.7, if we begin a trip at the even vertex C, either we become trapped at one of the odd vertices A and B or we return to C without traversing the entire network. However, by beginning our trip at one of the odd vertices (select A, for example), we can traverse all of the edges emanating from A by eventually leaving A and never returning. Furthermore, all of the edges leading to the other odd vertex B may be traversed by terminating our trip at B. Getting to and from the even vertices is, of course, no problem; but it is important not to terminate our trip at B before traversing all edges or we will become trapped there and will be unable to complete our trip. From our discussion, we may conclude correctly that, *if a network has exactly two odd vertices, it can be traversed by starting at either of the odd vertices and ending at the other.* The reader should experiment in traversing the network of Figure 9.2.7.

Since we cannot terminate a trip at more than one odd vertex in a network, *any network with more than two odd vertices is not traversable.* In fact, *in order to travel exactly once over each edge of a network with odd vertices, a separate trip is necessary for each pair of odd vertices.* Traveling exactly once over each edge of networks (a), (b), and (c) of Figure 9.2.5 requires 1, 2, and 3 trips respectively.

Finally, we return to the Königsberg bridge problem. Referring back to Figures 9.2.1 and 9.2.2, do you believe it is possible to cross each bridge but once on a single trip? If so, the network of bridges is traversable.

EXERCISE 9.2

1. Which of the following networks is traversable?

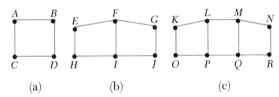

(a)   (b)   (c)

2. Referring to Problem 1, which vertices may be used as starting points in traversing the entire network in one trip:
   (a) For figure (a)?
   (b) For figure (b)?
   (c) For figure (c)?

3. In figure (c) of Problem 1, the network can be made traversable by removing only one edge. Which of the edges could be removed for this purpose? (There are four possibilities.)

4. (a) Is the following figure traversable? Explain.
   (b) At what points can one begin and end a trip in traversing the figure?

5. Following is a diagram of a cube. Its edges form a three-dimensional network.

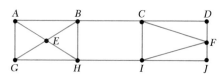

   (a) How many vertices does the cube have?
   (b) How many even vertices does the cube have?
   (c) How many odd vertices does the cube have?
   (d) How many distinct trips will it take to travel exactly once over each edge of the cube?
   (e) Draw a cube very lightly with your pencil. Then traverse its edges by marking the edges heavily and see if your answer to part (d) proves to be correct.

6. Referring to the Königsberg bridge problem (see Figures 9.2.1 and 9.2.2):
   (a) How many trips are necessary to cross each bridge exactly once? Explain.
   (b) How many bridges must be built to make a traversable network? Explain.
   (c) How many bridges must be built if a resident of Königsberg (no matter where in town he lives) is to be able to start at his home, cross each bridge only once, and end up at his home? Explain.

7. In the following diagram towns A, B, and C have been connected with roads in such a way that one can travel between any two towns without going through any other town. Furthermore, without the use of overpasses, no roads cross each other.

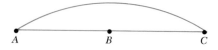

   (a) Can such a network of roads be built connecting four towns A, B, C, and D?
   (b) Can such a network of roads be built connecting five towns A, B, C, D, and E?

9.3 Euler's Formula for Polyhedra

We will be ready to consider Euler's formula for polyhedra after we develop a few additional concepts concerning plane connected networks. The connected network of Figure 9.3.1 has 5 vertices and 8 edges. There are also 4 closed regions that are determined by the network and labeled W, X, Y, and Z. These closed regions are called **faces**. While the faces are not part of the network itself, they are referred to as the faces of the network that encloses them. Each face is enclosed by a set of edges and vertices. Face W, for example, is enclosed by 4 edges and 4 vertices, while face Z is enclosed by 1 edge and 1 vertex. Each face is enclosed in such a way that any two points of the face may be connected by a path that does not cross an edge.

FIGURE 9.3.1

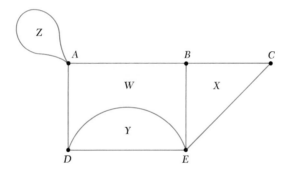

A formula relating the number of faces, vertices, and edges of any connected network is certainly not immediately apparent. However, if a connected network has a total of F faces, V vertices, and E edges, we propose to show that the following result is true:

$$F + V - E = 1.$$

In counting the faces, vertices, and edges of the network in Figure 9.3.1, we see that the above formula does indeed hold: $F + V - E = 4 + 5 - 8 = 1$.

In order to understand why the equation $F + V - E = 1$ is true for all connected networks, let us study the way in which any network can be constructed. By definition any network must have at least one edge, but it cannot have infinitely many edges. Consequently, any connected network may be built by starting with a single edge and annexing additional edges, one at a time, until the desired network has been built.

For the first edge of a network, there are only two choices:

(1) The edge has distinct end points.
(2) The end points of the edge coincide and we have a loop.

In either of the above cases, however, the formula $F + V - E = 1$ will hold (see Figure 9.3.2). Since our formula is true for a single edge whether or not its end points are distinct, we may begin a network with either kind of edge.

FIGURE 9.3.2

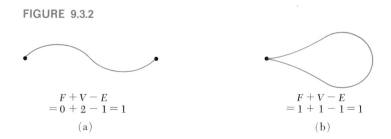

$$F + V - E$$
$$= 0 + 2 - 1 = 1$$
(a)

$$F + V - E$$
$$= 1 + 1 - 1 = 1$$
(b)

For illustrative purposes, we will arbitrarily begin a network with a loop (see (a) of Figure 9.3.3). Since our network is to be connected, each edge we annex must have one or both of its end points coincide with existing vertices (but edges may not intersect except at their end points). If only one end point of a new edge coincides with an existing vertex, we gain *1 vertex* and *1 edge,* and our formula continues to hold (see (b) of Figure 9.3.3). If, however, we annex an edge and both of its end points coincide with existing vertices, we gain *1 face* and *1 edge,* and our formula again is true. Furthermore, it makes no difference whether the end points of the new edge coincide with distinct existing vertices or with a single existing vertex (see (c) and (d) of Figure 9.3.3). Any further enlargement of our network will simply be a repetition of techniques already discussed, and the network will be enlarged by either a vertex and an edge, or by a face and an edge. Hence, the formula $F + V - E = 1$ will continue to be valid. Two more edges have been annexed to the network in (e) and (f) of Figure 9.3.3 for illustrative purposes, but no new principles are involved in so doing.

As soon as we define the terms *polygon* and *polyhedron,* we will be fully prepared to discuss Euler's formula for polyhedra. A **polygon** is a simple closed curve whose edges are straight line segments; a triangle, a rectangle, and a pentagon are examples. If each of the sides of a polygon are equal in length and each of the interior angles are equal in measure, then the polygon is said to be a *regular polygon;* a square is an example.

FIGURE 9.3.3

$F + V - E$
$= 1 + 1 - 1 = 1$
(a)

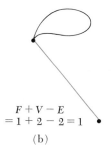

$F + V - E$
$= 1 + 2 - 2 = 1$
(b)

$F + V - E$
$= 2 + 2 - 3 = 1$
(c)

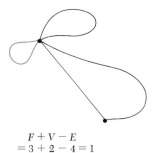

$F + V - E$
$= 3 + 2 - 4 = 1$
(d)

$F + V - E$
$= 3 + 3 - 5 = 1$
(e)

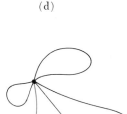

$F + V - E$
$= 4 + 3 - 6 = 1$
(f)

A **polyhedron** is a simple closed surface whose faces are bounded by polygons; a cube and a prism are examples. A polyhedron is a *regular polyhedron* if its faces are bounded by identical (congruent) regular polygons and the same number of edges emanate from each vertex; a cube, for example, is a regular polyhedron. There are only five regular polyhedra as pictured in Figure 9.3.4.

FIGURE 9.3.4

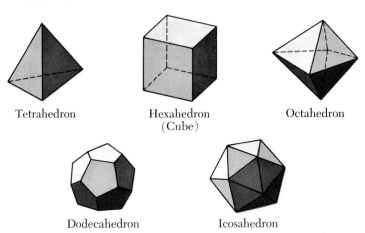

Tetrahedron   Hexahedron   Octahedron
                       (Cube)

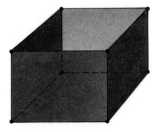

Dodecahedron   Icosahedron

In studying polyhedra, such as those pictured in Figure 9.3.4, René Descartes discovered as early as 1640 a formula relating the faces, vertices, and edges:

$$F + V - E = 2.$$

However, this formula for polyhedra was first proved in 1757 by Euler, and for this reason it is usually referred to as **Euler's formula for polyhedra**. Euler's formula can be used to prove that there are only five regular polyhedra.

Euler's formula for polyhedra is very similar to the formula for connected networks in a plane; the only difference is that the left member is equal to 2 rather than 1. The reason for this difference can be readily explained. If one face is removed from a polyhedron, it merely has one

FIGURE 9.3.5

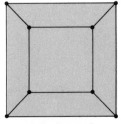

Cube with top
face removed

Cube distorted
into a plane connected
network with faces

less face—the number of edges and vertices remains the same. Having removed one face from a polyhedron, we obtain a figure with a "polygonal hole" in it that can be distorted by elastic motion and "flattened" into a plane connected network with faces. As an example, see Figure 9.3.5. We know, however, that $F + V - E = 1$ holds for a plane connected network. Hence the polyhedron, which has one more face than the corresponding connected network, must be such that $F + V - E = 2$.

EXERCISE 9.3

1. Examine the following connected network:

   (a) What is the number of faces?
   (b) What is the number of vertices?
   (c) What is the number of edges?
   (d) Verify that $F + V - E = 1$.

2. Which of the following figures are not connected networks?
   (a)
   (b)
   (c)
   (d)

3. Consider the following network:

   (a) How many edges emanate from vertex C?
   (b) If C is not considered to be a vertex, will the figure continue to represent a network?
   (c) If C is not considered to be a vertex, will the equation $F + V - E = 1$ hold?
   (d) How many edges emanate from D?
   (e) Must D be considered to be a vertex for the figure to be a network?

4. (a) Construct a connected network with 3 vertices and 3 faces. How many edges *must* the network have? Explain.
   (b) If a network has 4 vertices and 4 faces, how many edges must it have? Verify your prediction with a drawing.
   (c) If a network has the same number of vertices and faces, can it have an even number of edges? Explain.

5. The following circle has 6 edges and 6 vertices:

(a) Does $F + V - E = 1$ hold?
(b) If the circle had 3 vertices, how many edges would it have?
(c) If the circle had n vertices, how many edges would it have?

6. Refer to the octahedron of Figure 9.3.4:
   (a) What is the number of faces?
   (b) What is the number of vertices?
   (c) What is the number of edges?
   (d) Verify that $F + V - E = 2$.
   (e) Are the edges of the tetrahedron traversable? Explain.

7. Refer to the tetrahedron of Figure 9.3.4:
   (a) How many trips are necessary to traverse the edges of the tetrahedron? Explain.
   (b) Remove the base face and sketch the remaining portion of the tetrahedron after distorting and "flattening" it into a plane connected network with faces.
   (c) Using your figure of part (b), verify that $F + V - E = 1$.

9.4 Map Coloring or the Four-Color Problem

Leonard Euler, who solved the Königsberg bridge problem and proved the polyhedron formula relating faces, vertices, and edges, was one of the most prolific mathematicians in history. Although he lost sight in one eye at age 28 and became totally blind at 59, Euler continued to work until his death in 1783 at the age of 76. His many writings in mathematics as well as astronomy, physics, theology, oriental languages, and medicine will fill well over 50 large volumes. The task of publishing his complete works was begun in 1909 by a Swiss scientific society.

Having given full credit to the genius of Euler, one still may wonder if he could have solved a simply stated topological problem that was posed after his death by the German mathematician Möbius in 1840 and later (about 1850) in an independent manner by a British student named Francis Guthrie. While coloring a map of England, Guthrie began to wonder how many different colors would be necessary to color any number of regions in such a way that no two bordering regions would be of the same color.

Obviously, a checker board can be colored with only two colors; but what about a complex map, such as the states of the United States of America? One might imagine that a great many colors would be needed, but Guthrie could find no map requiring more than four colors. It is quite easy, however, to exhibit a map that does require four colors. Figure 9.4.1 is an example.

FIGURE 9.4.1

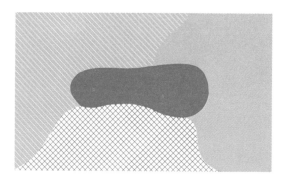

Since no map, however involved, has ever been demonstrated to require more than four colors, it is widely believed that in fact no map requires more than four colors; and the problem to either prove or disprove this conjecture has become known as the **four-color problem**.

Through Guthrie's brother the four-color problem was referred to the English mathematician Augustus De Morgan, who was unable to solve it, and the question was all but forgotten until Arthur Cayley announced at a meeting in 1878 that he had unsuccessfully attempted to find a solution to this topological problem. Since Cayley (1821–1895) was one of the most famous and respected mathematicians of the nineteenth century, his announcement attracted much attention. The year following Cayley's announcement a lawyer named A. B. Kempe announced that he had proved that no more than four colors are necessary in the coloring of any map. Over ten years after the four-color problem was apparently solved, a flaw was exposed in Kempe's proof by P. J. Heawood. However, Kempe's work was of some value, for Heawood showed that Kempe's techniques could be used in proving that every map can be colored with no more than *five* colors; and that at least was a step in the right direction.

The four-color problem continues to baffle the world's best topologists as well as many amateurs who have attempted to solve it. As of this writing, no general solution has been found.

Although the four-color problem is still an enigma, many interesting theorems concerning map coloring have been proved. Rather than presenting detailed proofs of theorems, however, we shall encourage the reader to formulate a few map coloring hypotheses for himself by asking him to work the exercises that follow.

   In discussing map coloring we will usually refer to the regions of a map as countries. The country boundaries of many ordinary maps are connected networks, which were studied in Section 9.2. In fact, a clever application of Euler's formula for connected networks in a plane, $F + V - E = 1$, is the key to the Kempe–Heawood proof that no map requires more than five colors.

EXERCISE 9.4

   1. Redraw the following maps and color them with as few colors as possible. How many colors are necessary for each?

   2. Which of the following map boundaries constitute a connected network?

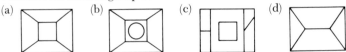

   3. Redraw the maps of Problem 2 and color them with as few colors as possible. How many colors are required for each?

   4. Color a map similar to each of the following with as few colors as possible. How many colors are required?

   5. Each of the following maps has only even vertices in its interior. Make a copy of each of them and color with as few different colors as possible.

   (a) Is it possible to color each map with only 2 colors?
   (b) Draw a rather complex map of your own having only even interior vertices. Can it be colored with only 2 colors?
   (c) Make a conjecture concerning the number of colors necessary to color any map that has only even interior vertices.

6. Draw a map with exactly 4 countries that requires for coloring:
   (a) No more than 2 colors.
   (b) 3 but no more than 3 colors.
   (c) 4 colors.

7. Attempt to draw a map with an odd interior vertex requiring only 2 colors. Is this possible?

8. Distort the following map into a topologically equivalent map such that each country is a polygonal surface. Keep the vertices arranged approximately as they are now and be certain that the correct number of edges emanate from each vertex.

   (a) Does the distorted polygonal map have the same number of regions as the original map?
   (b) How many interior vertices are there in each map?
   (c) Are interior vertices odd or even?
   (d) How many colors are necessary to color each of the maps?

9.5   One-Sided Surfaces

One is tempted to accept as axiomatic the notion that a surface must have two sides. We have been taught that a sheet of paper can be used to represent a surface; we have been taught that a sheet of paper has two sides; we are frequently requested to write on one side of a sheet of paper but not the other; and we would probably all do approximately the same thing if we were requested to color one side of a sheet of paper red and the other side green. But what criteria can we use in determining whether a surface has one or two sides?

Let us agree that if a piece of paper is used to represent a surface, then the surface it represents has only one side if we can draw a path connecting any two points A and B on the paper without crossing an edge. Since there are pairs of points A and B on an ordinary sheet of paper that cannot be connected by a path without crossing an edge, the paper represents a two-sided rather than a one-sided surface. Furthermore, the sheet of paper has only one edge since any two points on its edge may be connected by a path that does not cross its surface. Note that the edge is a simple closed curve.

Are there surfaces with only one side? Since surprises are not uncommon in topology, it should not be completely unexpected to discover that there are. The German mathematician Augustus F. Möbius (1790–1868) discovered in 1858 that a one-sided surface could be created merely by taking a strip of paper, giving it a half-twist, and gluing the ends together to form a ring. Such a ring, which is pictured in Figure 9.5.1, is appropriately called a **Möbius strip**. Since any two points on a Möbius strip may be connected without crossing an edge, it is, by definition, a one-sided surface. Even though two points on a Möbius strip may appear to be on opposite sides, they can always be connected by a path without crossing an edge and are, therefore, on the same side.

FIGURE 9.5.1

A Möbius strip

Not only is the Möbius strip a surface with but one side, it has only one edge. This is easily demonstrated by showing that any two of its edge points may be connected by a path that does not cross the surface. Further examination will prove the edge of the Möbius strip to be a closed curve topologically equivalent to a circle.

Although the properties of the Möbius strip are in themselves interesting, they have also been put to practical use. A Möbius strip conveyor belt will last longer since it has only one side. Möbius strip electronic resistors have useful properties, and Möbius recording tape will play twice as long without repetition since its entire surface can be used.

The exercises which follow request the reader to perform a few experiments, some of which produce rather unexpected results. In this way, the reader can discover for himself additional properties of one-sided surfaces.

EXERCISE 9.5

For the following problems you will need several strips of paper measuring approximately 2 × 10 inches, a pair of scissors, and some tape or quick-drying glue. Be certain to do a good job of taping so that no part of your model will fall apart after cutting.

1. Give an end of one of your strips of paper (as described in the directions) a half-twist (a twist of 180°) and tape it to form a ring similar to the Möbius strip of Figure 9.5.1.
   (a) Can you "travel" along the entire edge without crossing the surface?
   (b) How many edges does the ring have?
   (c) Could you by elastic motion distort the edge into a circle?
   (d) If an insect lands at some point on the surface, will it be able to crawl to any other point on the surface without crossing an edge?
   (e) Color one side of the paper as far as you can, but do not cross over the edge. How many sides has the surface?

2. Using a new strip of paper make another ring with a half-twist as described in Problem 1. Mark a large dot on the center of the surface and draw a center line parallel to the edge; continue until you return to the starting point. Cut along the center line and see what happens.
   (a) Color a side of the new ring. How many sides has the new ring?
   (b) How many half-twists are in the new ring?
   (c) How many edges has the new ring?
   (d) Cut along the center of the new ring parallel to its edges. How many rings do you have now?
   (e) How many half-twists are in each?
   (f) How many sides does each have?
   (g) How many edges does each have?

3. Start with a new strip and form a ring with one half-twist. Mark a large dot about one-fourth inch from the edge and draw a line parallel to the edge; continue until you return to the starting point. Cut along this line and see what happens.
   (a) How many rings result?
   (b) Describe each ring in terms of the number of half-twists, number of sides, and number of edges.

4. Start with a new ring with two half-twists. Cut it along the center.
   (a) How many rings result?
   (b) Describe each ring in terms of the number of half-twists, number of sides, and number of edges.

5. Using a 1 × 10 inch strip, make a ring with 3 half-twists. Cut it along the center.
   (a) What obvious property does the resulting ring have that previous rings did not have?
   (b) How many sides and edges does the ring have?

6. How many sides and edges has each of the following figures (a), (b), and (c)?

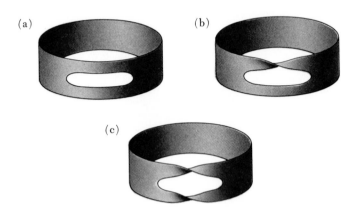

(a)   (b)

(c)

REVIEW EXERCISES

1. Sketch three figures that are topologically equivalent to the numeral 8.

2. Can a circle be distorted through elastic motion in such a way that it becomes:
   (a) a square?   (b) a triangle?
   (c) a trapezoid?   (d) the letter B?

3. Is a saucer topologically equivalent to:
   (a) a drinking glass?   (b) a sheet of paper?
   (c) a coffee cup?   (d) a spoon?

4. Construct a connected network in a plane having:
   (a) 2 faces, 4 vertices, and 5 edges.
   (b) 2 faces, 5 vertices, and 6 edges.
   (c) 2 faces, 1 vertex, and 2 edges.

5. Construct a connected network having:
   (a) 1 face, 1 vertex, and 1 edge.
   (b) 1 face, 2 vertices, and 2 edges.
   (c) 1 face, 3 vertices, and 3 edges.

6. Construct a connected network having:
   (a) 1 face, 1 vertex, and 1 edge.
   (b) 2 faces, 1 vertex, and 2 edges.
   (c) 3 faces, 1 vertex, and 3 edges.

7. (a) What generalization can you make concerning the number of vertices and edges in a plane connected network if it has exactly one face?
   (b) What generalization can you make concerning the number of faces and edges in a plane connected network if it has exactly one vertex?

8. Make a relatively simple map, whose boundaries constitute a connected network, that can be colored with no fewer than:
   (a) 2 colors.
   (b) 3 colors.
   (c) 4 colors.

9. Refer to Figure 9.2.2, which represents a schematic diagram of the Königsberg bridges. Suppose another bridge is built between points A and D. Can the network then be traversed? If so, explain how.

10. If a connected network has 40 odd vertices, how many distinct trips will it take to travel exactly once over each edge? Explain.

Paul Klee, *Schwankendes Gleichgewicht*
1922. Aquarell. 34.5 x 17.7 cm.
Paul Klee Foundation, Museum of Fine Arts, Berne

10 MATHEMATICAL MACHINES

10.1 Introduction

Most animals depend almost exclusively upon their own physical strength to move about and to obtain the food they need to survive. A few have learned to utilize, in a limited way, other forms and sources of energy. The buzzard, for example, by taking advantage of up-drafts of air currents, may sail in the air for hours without flapping its wings.

Man, however, has extended his physical prowess beyond all other animals—not because he has the largest muscles of any animal, but because he has invented mechanical tools, machinery, and other devices that respond to his will and are powered by various sources of energy independent of his own physical strength. He has used energy from the wind to sail the seas, energy from the waters of the rivers and from coal to drive turbines and generate electricity, energy from petroleum to drive automobiles and propel jet aircraft. Finally, he has learned to produce energy from the atom in such quantities that he must exercise considerable judgment in its use or run the risk of destroying nearly all forms of life on earth, including himself.

Not only has man extended his physical powers by using various sources of energy, he has also extended his hearing by building radios and

235

telephones and his sight by constructing microscopes, telescopes, and television sets. For a variety of reasons and purposes, man has invented many kinds of tools and machines, but all were made possible by the thinking power of his brain.

Finally, we ask the question, "Can man build tools and machines to extend his thinking power?" In a sense mathematics is such a tool; but in using mathematics it frequently becomes too time consuming to perform the operations necessary to solve a problem. Even if man determines the operations that are necessary to solve a problem, this may be of little help if it takes years to perform the manipulations. Nearly everyone knows, however, that man has developed computers to perform mathematical manipulations at high speeds.

The development of mathematical computers is an interesting chapter in the history of mankind, for not only has the computer enabled man to multiply his thinking power, but it has also enabled him to do what previous generations considered all but impossible—fly to the moon.

Can the modern sophisticated electronic computer actually think? The answer to this question probably depends upon how one defines the word "think." In any event we must concede that the computer has become an almost indispensable servant to the mind of man.

10.2   Kinds of Computers

Computers may be classified into two basic groups: the analog computers and the digital computers. Analog computers are designed and built by applying the concept of *measurement,* while digital computers use the concept of *counting.*

A thermometer is an example of a simple analog computer. The height of the liquid in the tube is used as a measure of the temperature. There are many other examples of simple analog devices. The physical distance the dial moves on a bathroom scale is used as a measure of your weight. The distances the hands on a watch move are used to measure time. The distance the needle on a speedometer moves from its original position is used to measure speed. The slide rule uses distances to correspond to the products, quotients, roots, and powers of numbers; it too is an example of a relatively simple analog computer. More sophisticated analog computers may measure voltages or resistances rather than distances or rotations, but the principle of measurement is nevertheless basic in their structure.

Since the analog computer operates on the principle of measurement and our ability to measure is always subject to error, the analog computer has limitations when a high degree of accuracy is desired. However, analog computers are relatively inexpensive to build and are quite versatile in their applications. We will probably continue to manufacture thousands of thermometers, speedometers, thermostats, watches, and so on. Even the gearing mechanism between the rear wheels of an automobile (called the *differential*) is a kind of analog computer. In turning a corner, for example, any extra speed gained by one wheel is subtracted from the other.

To further illustrate the measurement principle of the analog computer, let us imagine we wish to find the simple sum $3 + 5$. Using two rulers, put the beginning of the second ruler at the location of 3 on the first ruler. Then look at the numeral 5 on the second ruler and see what numeral it corresponds to on the first ruler (see Figure 10.2.1). The 5, of course, will be near the 8 and we conclude that $3 + 5 = 8$. However, it may well be that the 5 is closer to 8.1 than it is to 8.0. If a reading of 8.1 is taken, we then have a small error in our sum. In a problem with many steps in its solution, such errors may accumulate and the final result will be quite far from the correct result.

FIGURE 10.2.1

Consider once again the problem of finding the sum $3 + 5$. Suppose we raise 3 fingers on one hand and 5 fingers on the other, and then count the total number of raised fingers. We will, of course, obtain exactly 8, never 8.1 or 7.9. In finding the sum in this way, we have used our hands as a simple *digital computer* since the principle of counting was involved. An ordinary desk calculator is a digital computer; a cash register is a digital computer; and most of the modern electronic computers are digital. One of the earliest of the digital type computers is the Chinese abacus, which was developed thousands of years ago. We will have more to say about the abacus in the next section.

Although the remainder of this chapter will deal primarily with digital computers, let us emphasize that analog devices are important. A fair degree of success has been achieved recently in combining some of the best features of the analog computers and the digital computers into *hybrid* machines. In fact, much more hybridization is anticipated.

EXERCISE 10.2

1. How does man compensate for the fact that the physical strength of many animals exceeds his own?

2. In what way does a computer assist man in his mental work?

3. Name and briefly describe the two basic types of computers and give three simple examples of each.

4. (a) Make a diagram showing how two rulers can be used to find the sum 4 + 9.

(b) Make a diagram showing how two rulers can be used to find the difference 10 − 3.

(c) What type of computing device is involved in using two rulers for addition and subtraction? Explain.

10.3   Computing Machines of Early Origin

If we exclude using fingers (and toes) for computing, then the oldest known computing machine of any importance is the abacus. Although the inventor is not known, early forms of the abacus probably appeared about 3000 B.C. Over the years the abacus has changed but little in form, and it proved so successful that it is still used in some countries, such as China, Japan, India, and Russia.

The success of the abacus as a computing device is based upon its simplicity of construction and operation. Figure 10.3.1 shows a sketch of a Chinese abacus. Beads strung on wires or rods are used as counters to represent numbers. Those on one line represent units; on the next line, tens; on the next, hundreds; then thousands, and so on. In the Chinese abacus each wire is partitioned in such a way that five beads are on one side of the partition and two are on the other. Each of the five-bead group counts as a single unit while each of the two-bead group counts as five units. The abacus of Figure 10.3.1 is set to represent the number 18,462. If a number is to be added to 18,462, more beads may be moved toward the center

FIGURE 10.3.1

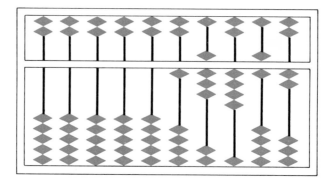

partition. If a number is to be subtracted from 18,462, beads may be moved away from the center partition.

In using the abacus, beads representing ten units on any given wire may be returned to their original positions provided one bead of the next higher order is moved toward the center partition. With a little practice it is not particularly difficult to learn to add and subtract on the abacus; multiplication and division require a somewhat higher degree of skill. Moreover, many short-cuts are possible with the abacus, and mental arithmetic can be used in conjunction with the movement of the beads. A distinct disadvantage of the abacus is the fact that each step in a problem is automatically erased when the next operation is performed. Hence, it is impossible to check a problem without reworking it; and, if an error is made, one must start again at the beginning of the solution.

In the United States the abacus serves primarily as a teaching aid in arithmetic, and we are inclined to consider its use in other respects as quite primitive. However, in the hands of a skilled operator, arithmetic computations can be performed quite rapidly.

During the American occupation of Japan following World War II, a contest was arranged to demonstrate the superiority of the electric calculating machine (costing $700) over the abacus (costing 25¢). Over 3000 spectators gathered to watch the match between 22-year-old Kiyoski Matsuzaki and 22-year-old Thomas Ian Wood. Matsuzaki was a Japanese Communications Ministry clerk with special training in the use of the abacus; Pvt. Wood was an Army finance clerk with four years' experience on modern machines. Matsuzaki won the contests in addition, subtraction, and division; he also won the final contest in the solution of a composite problem. Only in multiplication did Pvt. Wood surpass Matsuzaki. The contest apparently did a great deal to create a respect on the part of the Americans for those of another culture.

Another computing device which has been employed for many years is the slide rule. While the abacus is a simple digital computer whose function depends upon the principle of counting, the slide rule is a simple analog computer whose function and use is based upon the principle of measurement. Since a discussion of exponents and logarithms is necessary for an adequate treatment of the slide rule's construction, we will satisfy ourselves at this time by saying that the slide rule is an inexpensive and handy device for finding products, quotients, powers, and roots of numbers. (Any mathematics or engineering student would probably be happy to discuss with the reader the use of the slide rule!)

The principle of the slide rule was conceived by the Englishman Edmund Gunter in 1620, but credit for building the first slide rule similar to those in use today goes to William Oughtred (1574–1660). Oughtred's original slide rule was fabricated in about 1622.

Precursors of the modern electronic computers date back to the seventeenth century. In 1642, at the age of nineteen, the French mathematician Pascal constructed the first calculating machine that might be considered a prototype of those we know today. Pascal's computer (see Figure 10.3.2) could be used to add and subtract by turning telephone-like dials. Its principle of operation, which involved a system of gears, was similar to that of a present-day automobile odometer.

FIGURE 10.3.2

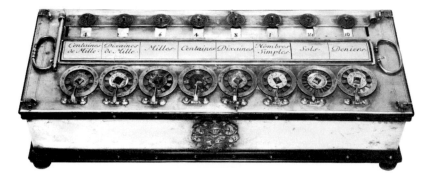

PASCAL'S COMPUTER

In 1671 the German mathematician Gottfried Leibniz (1646–1716) (one of the independent inventors of calculus), designed a machine for adding, subtracting, multiplying, dividing, and finding square roots. However, the machine itself was not built until 1694 (see Figure 10.3.3). Leibniz expressed the opinion that excellent men should be doing fruitful work rather than wasting their time like slaves in making the calculations neces-

FIGURE 10.3.3

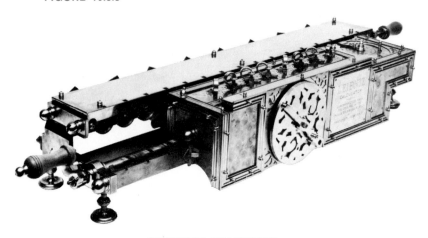

LEIBNIZ'S COMPUTER

sary for such things as mathematical tables. Although Leibniz's machine was not very reliable and never became widely used, it did express the frustration of the mathematician who is forced to make the necessary but boring computations when solving certain kinds of problems. Perhaps his machine also served as an inspiration for others to attack the problem of building even better computers.

One of the noblest attempts to build a highly productive computer was made by the English professor and mathematician Charles Babbage (1792–1871). Although he never completed the model envisioned, in 1822 Babbage did build an abbreviated working model called the "Difference Engine" (see Figure 10.3.4). The model could generate logarithmic and astronomical tables accurate to six decimal places. For this he received the first Gold Medal of the Royal Astronomical Society.

FIGURE 10.3.4

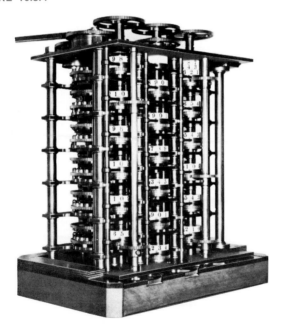

BABBAGE'S COMPUTER

Since electricity was not available at that time, the machine that Babbage envisioned was to be powered by steam. It was planned that data and operating instructions would be fed into the machine by the use of punched cards. The machine would be able to calculate results accurate to 20 digits; it would be able to store up to 1000 50-digit numbers which could be retrieved at a later date; and it would be able to modify its own instructions according to previously computed results. If Babbage could have completed his machine as planned, he would have had a digital computer comparing favorably with those produced many years later. It eventually became clear, however, that his plans were too grandiose for the technology of his time, and that the approximately 50,000 parts necessary for his computer could not be made with the required precision. For this and other reasons, funds for his project that had been forthcoming from the British government were terminated. Had Babbage been born a century later, his dreams could easily have been fulfilled.

Babbage's life is somewhat reminiscent of that of Leonardo da Vinci. While Babbage designed computers that could not be built during his life because of a lag in technology, da Vinci designed aircraft similar in principle to the first airplanes and helicopters built during the twentieth century. (It is interesting that the International Business Machines Corporation has built many models from the various drawings of machines by da Vinci.)

Although Babbage invested much of his life as well as much of his own money in the building of a sophisticated computer, he finally realized that he would probably never complete the project. Before his death he wrote, "If, unwarned by my example, any man shall succeed in constructing an engine embodying in itself the whole of the executive department of mathematical analysis . . . I have no fear of leaving my reputation in his charge, for he alone will be able to fully appreciate the nature of my efforts and the value of their results."

EXERCISE 10.3

1. (a) What type of computer is the abacus?
   (b) Briefly describe its construction and operation.
2. Each of the following devices may be thought of as a computer. Label each as either an analog device or a digital device.
   (a) yard stick   (b) tally marks
   (c) spring scales   (d) mercury thermometer
   (e) barometer   (f) speedometer
   (g) cash register   (h) gasoline gauge
   (i) poker chips   (j) hour glass
3. Briefly describe the contributions of Pascal and Leibniz to the development of computers.
4. (a) Explain why Babbage was "born too soon."
   (b) What contribution did Babbage make to computer technology?

10.4 Modern Electronic Computers

The development of the precision machine tool industry and the electronics industry have made possible the building of modern electronic computers.

The first person to use electrical tabulating equipment to analyze statistical data was Dr. Herman Hollerith of the U. S. Census Office. In 1890, when faced with the monumental task of tabulating and organizing the census data, Hollerith developed a code system whereby each person's name, address, sex, age, occupation, and other vital statistics could be represented by holes punched in paper cards. According to the positions of the holes, the cards could be sorted by electromagnetic relays activated by contacts made through the holes. As a result, even though there was more data to process in the 1890 census than in the 1880 census, the job was completed in much less time.

By expanding on the concepts developed by Hollerith, faster machines were built for handling accounting problems involving sales, payrolls, insurance policies, billings, and so on. Punched-card equipment was particularly helpful in processing the vast amount of data that had to be recorded as a result of the Social Security Act passed by Congress in 1935.

As businessmen, mathematicians, and scientists collected more and more data and posed new problems, the need for more versatile and much faster operating machines became clearly evident. By the use of various kinds of electrical and mechanical devices, including electro-mechanical relays, better machines were constructed. However, the fact that computers had mechanically moving parts severely limited the speed with which the machines could operate. While the electricity that activated the moving parts could travel at about 186,000 miles per second, it was obvious that no such speed could be expected from a mechanical motion.

The internal mechanical motions within the computer were eventually eliminated by the use of *vacuum tubes*. In considering the way in which vacuum tubes can be used in a digital computer, it is important to know that mathematical computations made by the machine are performed in a binary (base two) numeral system. By using a binary system, any number may be represented by a sequence of only two digits, 0 and 1 (see Section 4.6). Furthermore, various sequences of zeros and ones may be coded to represent letters of the alphabet, words, or almost anything we might choose. Since an electrical circuit in a computer is either open or closed, two alternatives are automatically present to represent the digits 0 and 1. The logic within a computer is the same whether the digits 0 and 1 are indicated by the presence or absence of a hole in a paper card or in some other way. The only requirement is that something within the machine have two states which can be used to represent the digits 0 and 1. An electro-mechanical relay that is either open or closed was such a device, but, as mentioned previously, the mechanical motion associated with its operation was a time-consuming factor. A vacuum tube, however, can be used to represent the digits 0 and 1 without mechanical motion. This may be done merely by placing the tube in a conducting or nonconducting state. Since a vacuum tube may be placed in a conducting or a nonconducting state in about a millionth of a second without mechanical motion, the use of vacuum tubes in computers was a giant step forward.

The first all-electronic computer, named ENIAC, was completed in 1946. ENIAC used 19,000 vacuum tubes, occupied 1500 square feet of floor space, weighed 30 tons, and used enormous quantities of electricity (enough to light 2000 100-watt light bulbs). ENIAC was developed by graduate student J. Presper Eckert and physicist Dr. John W. Mauchly at the University of Pennsylvania. The speed of ENIAC may be indicated by the fact that it could perform 5000 additions or 350 multiplications per second.

From 1946 to the present, progress in the design of computers has been rapid. One important change has been the replacement of vaccum tubes by *transistors*. While vacuum tubes are relatively large, costly, and create great amounts of heat, transistors are quite small, economical to build and operate, create almost no heat, have a long life, and are highly reliable. Much work has recently been done on computers to miniaturize their parts, to increase their reliability and versatility, and to make them easier to operate at a smaller cost. The use of *integrated circuits*, which are very compact (see Figure 10.4.1), has done much to accomplish these objectives. Some relatively inexpensive but versatile computers are now smaller than typewriters.

Although the computers of today can perform up to 3,000,000 or more operations per second, even faster operation is required in the solution of certain kinds of problems. For example, great speed is needed to analyze what happens when an atomic reaction occurs. Recently a high-speed computer named ILLIAC IV was designed at the University of Illinois under a grant from the Department of Defense. It has been built by Burroughs Corporation and is now installed at the Ames Research Center of the National Aeronautics and Space Administration in California. ILLIAC IV has 64 processing units designed to operate simultaneously on portions of the same problem. With this arrangement, 100,000,000 instructions per second can be performed, thus enabling certain kinds of problems to be solved many times faster than was previously possible. Figure 10.4.2 is a picture of ILLIAC IV.

An important property of modern computers is their ability to store operating instructions (a *program*) as well as data. Although discussed earlier by Babbage, the concept of a stored-program computer was proposed and developed in the 1940's by Dr. John von Neumann. Not only can modern computers store operating instructions and other data in their memories, but they can also modify their instructions according to previous computations and results. In a sense, they can "learn" from their experiences.

A computer may have information stored externally in the form of cards with punched holes, or in the form of characters printed with magnetic ink. Most people have seen the "please do not fold, spindle, or mutilate" punched cards. Nearly all bank checks now have a person's

FIGURE 10.4.1

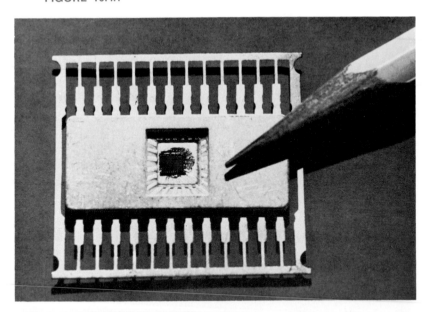

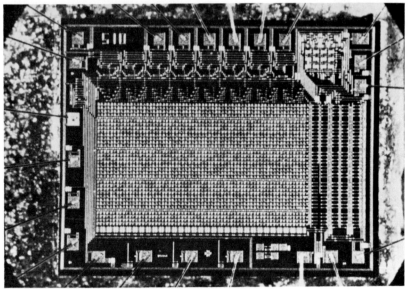

INTEGRATED MEMORY CHIP.

The central portion of the upper picture shows an integrated memory chip, the edge of which is about the length of a pencil point. This chip (enlarged in the lower picture) contains about 2500 electronic devices.

FIGURE 10.4.2

ILLIAC IV.
ILLIAC IV is a giant, parallel-processing computer
which can perform 100,000,000 operations per second.
It is used for world-wide weather forecasting and other
highly complex data processing applications.

account number printed on each check with magnetic ink. Also, if you
examine a cancelled check, you will probably notice that the amount of
the check has been printed at the bottom in magnetic ink. Another way
of storing computer data externally is by using punched paper tape or
magnetic tape.

Computers can store information internally on *magnetic disks*, which
are somewhat like high-fidelity records in appearance; *magnetic drums* are
also used. One of the best ways that has been invented for storing information so that it may be quickly located involves the use of *magnetic cores*.
A core is shaped like a doughnut and is about the size of a pin head (see

FIGURE 10.4.3

MAGNETIC CORES (enlarged)

Figure 10.4.3). Each core can be used to represent either the digit 0 or the digit 1 according to whether it is magnetized in a clockwise or counterclockwise direction. Since the magnetic cores are so small, thousands or even millions of them may be placed in a single computer without taking up a great deal of space. Whether a computer uses magnetic disks, magnetic drums, or core storage, that part of the computer used for storing information is called its *memory*. The functions of the memory and other principal parts of a computer are discussed in the next section.

EXERCISE 10.4

1. Briefly relate Hollerith's contribution to data processing.

2. Why is the study of a base two numeral system of great significance in understanding the functioning of digital computers?

3. Why was the use of vacuum tubes a big step forward in computer technology?

4. Briefly describe the significance of ENIAC as related to computer technology.

5. In what ways is the transistor superior to the vacuum tube in computer construction?

6. What contribution did Dr. John von Neumann make to computer technology?

7. What is meant by a computer's *memory?*

8. Without haste the author found the value of $\frac{(2.01)^2(427.23)}{82}$ in less than 15 seconds on a small computer (the Monroe 1655). For comparison, make the computation using pencil and paper. What was your result? How long did it take?

10.5 How a Digital Computer Works

Since even experts in the computing field are experiencing difficulty keeping up with the latest developments in computer technology, it is obvious that any brief discussion of how a computer works is necessarily an oversimplification. Nevertheless, we can make a few meaningful generalizations concerning the design and operation of a digital computer.

The operation of a computer involves three steps:

(1) **Input**: The computer accepts data and a set of instructions concerning what is to be done with the data in order that a final result can be produced.

(2) **Processing**: Operations are performed on the data according to instructions.

(3) **Output**: The final result exits from the computer.

The *input* may be entered into the computer in a variety of ways including punched cards, punched paper tape, and magnetic tape. See Figure 10.5.1. Each of these devices for input has some binary property. Certain locations on the paper cards or on the paper tape are either punched or not punched, and certain locations on the magnetic tape are either magnetized or not magnetized (or are magnetized in one of two different directions). The magnetized spots on magnetic tape are quite small and close together. Consequently a $10\frac{1}{2}$-inch roll of tape may contain as much information as 250,000 punched cards. High-speed computers can read 120,000 characters per second from tape, which is about 100 times faster than reading punched cards. Furthermore, information can be stored almost indefinitely on tape, or it can be erased and the tape used for recording new information.

Since the output of a computer can only be as good as the input, it is important that the input contain accurate data as well as an accurate set of instructions for handling the data. The set of instructions fed into the computer is called a *program*, and the person preparing the instructions

FIGURE 10.5.1

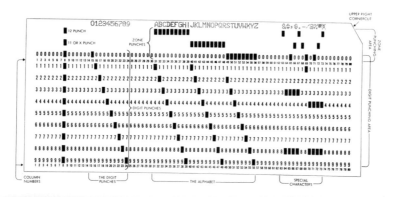

PUNCHED CARD, PUNCHED TAPE, AND MAGNETIC TAPE

is called a *programmer*. Almost anyone can learn to program a computer to process simple data or to work simple arithmetic problems. Preparing a program for business data-processing, such as that which is used in the preparation of a payroll, requires little specific mathematical training. However, a knowledge of higher mathematics is required to program a machine to solve complex mathematical or scientific problems. Since a computer functions on the basis of logic, a good programmer must have the ability to reason logically; if he also has a high degree of ingenuity coupled with a vivid imagination, he is likely to become a respected and sought-after programmer. A programmer, of course, needs to know what kind of language the computer understands so that he can write the program in a form that the computer can "read." Much work has been done recently to make computers that are better "readers." The ultimate, of course, is to build a machine that can read ordinary exposition or listen to a human voice and do all of its own translating into binary form. Some success has already been achieved in this respect. In summary, we might say that the function of the input unit of a computer is to accept and convert man-readable instructions and data into a machine-readable electrical code which the central processing division of the machine can use.

The *central processing unit* of a computer contains three distinct but connected divisions that work together in processing the input data in accordance with the programmed instructions. These divisions are called the *control unit*, the *memory unit*, and the *arithmetic unit*. See Figure 10.5.2. The control unit sees that everything happens in the correct sequence and it is, in a sense, the "supervisor" of the entire processing division. Control can transfer information from one unit of the computer to another as

FIGURE 10.5.2

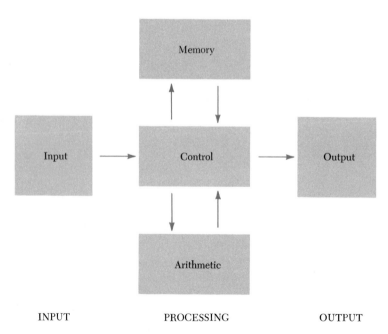

SCHEMATIC DIAGRAM OF A DIGITAL COMPUTER

necessary. The memory unit is used for storing data, operating instructions, intermediate results, and so forth. The arithmetic unit performs all necessary computations such as addition and subtraction or whatever other manipulations are desired. Control, of course, transfers from memory to arithmetic all data, instructions, intermediate results, and so on that arithmetic may need for any of its operations.

After the input data has been processed according to the program, it goes to the *output unit.* Output may be in the form of punched cards, punched paper tape, magnetic tape, printed material, or even in audio form. High-speed printers have proved to be one of the most useful means of getting information out of a computer. A high-speed printer can print 1200 lines a minute, which is as much typing as a good typist could be expected to do in several hours.

10.6   Applications and Implications for the Future

Recent developments in computer technology have been little less than phenomenal. In 1951 the first commercial computer was delivered for use. In 1970 there were probably over 100,000 computers in the world and the number is rapidly increasing. Through various terminal devices, many people in widely scattered locations may now use the same computer at the same time. Thus the number of people having access to computers is much larger than the number of computers.

Modern computers can tabulate almost unlimited quantities of information, and they can perform millions of operations per second in the solution of mathematical and other problems. Computers almost never make mistakes and they do not get tired.

Computers are able to store almost any type of information. They can maintain complete up-to-date records of census statistics, airplane and hotel reservations, income tax data, traffic violations, crimes committed, factory inventories, stolen cars, and so on. Furthermore, the computer can be used in conjunction with other machines to automatically print bills, refine petroleum, pilot aircraft, guide space ships to the moon, make cookies, mix concrete, run transit systems, register bank deposits and withdrawals, and build other machines. Computers have even been programmed to play chess and compose music. By simulating physical and chemical conditions, computers can be used to predict such things as whether or not an aircraft will fly properly or what chemical reactions will take place under given conditions. Applications of the computer to areas such as chromosome analysis, brain-wave analysis, and other biological, chemical, and physical phenomena are becoming commonplace.

It now seems possible to program a computer to run other machines that will be able to manufacture almost any product from a pencil to an automobile without human labor. But all this change is not without its problems. Will we be faced with an unsolvable unemployment situation as machines replace people? Or can those who are displaced by machines find other jobs that may be more stimulating? Perhaps the monotonous assembly line jobs will be replaced by more interesting activities such as exploring outer space, doing medical research, and so on. Must we learn to live with more leisure time and modify the not uncommon concept that virtue is directly proportional to how long and hard one works? Will the leisure-time industries be the growth industries of the future?

A particular situation that could prove to be problematical concerns the invasions of privacy by government and other interests. Until recently, records of a person's life were widely dispersed and it was relatively easy for someone to bury the past and start a new life. Are we now to have a central government file on each person with complete data on everything from his income, age, marital status, and occupation to his ethnic origin, religion, and sex habits? Will uses to which computers are put build up a distrust of the government and a resentment of being a dehumanized number? Is the unlawful destruction of computer data banks a specter of the future? The extent of concern for the unauthorized access to and use of data in computers is indicated by the fact that IBM announced plans to spend $40,000,000 over a five-year period to develop a system to block such access.

It seems that man must choose whether he will build automobiles or tanks, commercial aircraft or bombers, nuclear power plants or nuclear bombs. Undoubtedly, the computer can be either a monster or a servant to man. It is the responsibility of man to use the computer wisely, for it is he who builds the computer, poses problems for it to solve, and writes its programs. In any event, the future should be interesting.

EXERCISE 10.6

1. With reference to a computer, describe briefly what is meant by *input, processing,* and *output.*

2. Name the three divisions of the central processing unit of a computer and describe briefly the function of each.

3. It is believed that computers could be used to run most factory machinery automatically, thus eliminating the need for human labor. Comment on the statement: Eventually the computer will create an unsolvable employment problem.

4. It has been estimated that a 20-page dossier on every American could be stored on less than 5000 feet of computer tape. With this in mind, comment on the statement: A central government file on each person is a dangerous invasion of privacy and should be prohibited by law.

ANSWERS TO SELECTED EXERCISES

EXERCISE 1.3

1. (a) "Today is Monday"; "today is Wednesday." (c) "You will pay the rent"; "I will pay the rent."
2. (a) False. (c) True.
3. The connectives *and* and *or* are used to connect *two* statements to form a compound statement, while *not* is connected to only one statement.
4. (a)

P	$\sim P$	$\sim\sim P$
T	F	T
F	T	F

   (c) The truth values of $\sim\sim\sim P$ and $\sim P$ are identical. (Table omitted here.)
5. (a) Joyce is tall. (c) No mammals can swim.
6. (a) $H \wedge \sim D$ (c) $\sim(\sim H \wedge \sim D)$
7. (a)

P	Q	$\sim P$	$\sim P \vee Q$
T	T	F	T
T	F	F	F
F	T	T	T
F	F	T	T

   (c)

P	Q	$P \vee Q$	$\sim(P \vee Q)$
T	T	T	F
T	F	T	F
F	T	T	F
F	F	F	T

   (e)

P	Q	$P \wedge Q$	$(P \wedge Q) \vee P$
T	T	T	T
T	F	F	T
F	T	F	F
F	F	F	F

8. (a) $\sim(\sim G \wedge \sim R)$ (c) The statement is true if either George or Robert or both like poetry.

EXERCISE 1.4

1. See Figure 1.4.2.
2. (a) $M \wedge J$ (c) $M \vee \sim J$ (e) $\sim J \to \sim M$ (g) $J \leftrightarrow \sim M$

253

3. (a)

P	Q	$\sim Q$	$P \to \sim Q$
T	T	F	F
T	F	T	T
F	T	F	T
F	F	T	T

(c)

P	Q	$P \vee Q$	$(P \vee Q) \to Q$
T	T	T	T
T	F	T	F
F	T	T	T
F	F	F	T

(e)

P	Q	$P \wedge Q$	$P \vee Q$	$(P \wedge Q) \to (P \vee Q)$
T	T	T	T	T
T	F	F	T	T
F	T	F	T	T
F	F	F	F	T

4. (a)

M	R	$\sim R$	$M \leftrightarrow \sim R$
T	T	F	F
T	F	T	T
F	T	F	T
F	F	T	F

5. (a) Only when P is true.

6. (a)

P	$\sim P$	$P \to \sim P$
T	F	F
F	T	T

7. (a)

P	$\sim P$	$P \leftrightarrow \sim P$
T	F	F
F	T	F

EXERCISE 1.5

2. (a) *Converse:* If you can run fast, then you have long legs. *Inverse:* If you do not have long legs, then you cannot run fast. *Contrapositive:* If you cannot run fast, then you do not have long legs.

3. (a) and (d).

4. No. See Figure 1.5.1.

5. (There is no unique correct answer. For guidance refer to Figure 1.5.1. However, note that (d) has no false converse.)

6. (a) No. If an implication is true, its inverse may be either true or false.
7. (a) They have the same truth values.
8. (a)

P	S	$P \leftrightarrow S$
T	T	T
T	F	F
F	T	F
F	F	T

(c) They are identical.

9. (a)

P	Q	$\sim P$	$\sim Q$	$\sim (P \wedge Q)$	$\sim P \vee \sim Q$	$\sim (P \wedge Q) \leftrightarrow (\sim P \vee \sim Q)$
T	T	F	F	F	F	T
T	F	F	T	T	T	T
F	T	T	F	T	T	T
F	F	T	T	T	T	T

EXERCISE 1.6

1. (a) See Figure 1.6.1. (c) See Figure 1.6.3.
2. (a)

P	Q	$P \to Q$	$\sim P$	$(P \to Q) \wedge \sim P$	$\sim Q$	$[(P \to Q) \wedge \sim P] \to \sim Q$
T	T	T	F	F	F	T
T	F	F	F	F	T	T
F	T	T	T	T	F	F
F	F	T	T	T	T	T

(c) No. From the third and fourth lines of part (a) we see that when the premises of this type of implication are true, the implication itself may be either false or true.

3. (a) Valid; modus ponens. (c) Invalid. (e) Invalid.
4. (a) Modus ponens. (c) Chain rule. (e) Chain rule.
5. (a)

P	Q	$P \to Q$	$Q \to P$	$(P \to Q) \vee (Q \to P)$
T	T	T	T	T
T	F	F	T	T
F	T	T	F	T
F	F	T	T	T

(c) Statement (2), since statement (2) is always true if you're going to Mars is false.

6. (a)

K	L	$K \vee L$	$\sim K$	$(K \vee L) \wedge \sim K$	$[(K \vee L) \wedge \sim K] \to L$
T	T	T	F	F	T
T	F	T	F	F	T
F	T	T	T	T	T
F	F	F	T	F	T

Conclusion is valid.

(c)

L	M	$\sim L$	$L \to M$	$\sim L \vee M$	$(L \to M) \to (\sim L \vee M)$
T	T	F	T	T	T
T	F	F	F	F	T
F	T	T	T	T	T
F	F	T	T	T	T

Conclusion is valid.

EXERCISE 1.8

1. (a) The sum of any two even whole numbers is an even whole number. (b) By the definition of even numbers, $2a$ and $2b$ represent any two even numbers. Using the distributive property, the sum $2a + 2b = 2(a + b)$. The closure property of addition gives us $2(a + b) = 2c$. Hence $2a + 2b = 2c$, which is even since it has a factor of 2.
2. (a) The product of any two even whole numbers is an even whole number.
3. (a) Let $2a + 1$ and $2b + 1$ be any two odd numbers. Then their sum $(2a + 1) + (2b + 1) = 2a + 2b + 2 = 2(a + b + 1)$. But $2(a + b + 1) = 2c$, which is an even number. (Reasons for steps are omitted here.)
4. (Simply find the distinct factors of each number and count them.)
5. (a) 2 is a prime number, and it is also even.
6. (a) A whale is a mammal, and it lives in the sea.
7. Assume John did break the window. This assumption implies he must have been at Mrs. Arnold's house at 10:15 a.m. But John was in school at this time. Hence, our assumption that John broke the window must be false and we conclude that he did not break the window.

EXERCISE 2.3

1. There is no unique correct answer to this question. However, most new ideas seem to be met with their share of resistance.
2. The sets identified in (b) and (c) are well-defined.
3. (a) $\{5, 4, 3\}$; $\{5, 3, 4\}$; $\{4, 5, 3\}$; $\{4, 3, 5\}$; $\{3, 5, 4\}$; and $\{3, 4, 5\}$. (c) $\{9\}$
4. (a) A is equal to the set whose elements are 8, 9, and 10. (c) The set of all elements x such that x is equal to a whole number. (e) B is equal to the set of all x such that x is an element of set A and x is not equal to 8. (g) 5 is not an element of set K.
5. (a) $C = \{5, 10, 15\}$ (c) $7 \notin C$ (e) $S = \{y \mid y$ is an odd number less than $7\}$

6. (a) $\{x \mid x$ is an integer$\}$ (c) $\{x \mid x$ is an even number greater than 3$\}$
   (e) $\{x \mid x$ is a letter of our alphabet coming after $r\}$

EXERCISE 2.5

1. (a) See Definition 2.4.1.   (c) See Definition 2.5.1.
2. (a) A is a proper subset of B.   (c) The set whose elements are a and b is a subset of the set whose elements are a and b (itself).   (e) The empty set is a subset of the set whose elements are a and b.   (g) The empty set is not a proper subset of the empty set (itself).
3. A, C, and D.
4. (a) Yes. (b) No. (c) Yes.
5. $\varnothing$ represents the empty set, which has no elements; $\{\varnothing\}$ represents a set with a single element, the empty set.
6. B, D, E, G, H, and J.
7. B, E, G, and J.
8. (a) 4   (c) 2   (e) 0
9. The sets named in (a) and (d) are equal.

EXERCISE 2.7

1. See Definition 2.6.1.
2. See Definition 2.6.2.
3. Two sets are equivalent if their elements may be placed in a one-to-one correspondence, but equal sets must contain exactly the same elements.
4. (a) Yes.
5. (a) A, B, and D.
6. (a) $3 \times 2 \times 1 = 6$.
7. (a) Yes.   (c) Yes.
8. See Definition 2.6.3.
9. (The procedure will show that there are 5 elements in the set.)
10. (a) $\{1, 2, 3\}$; 3   (c) $\{1, 2, 3, 4, 5\}$; 5   (e) $\{1, 2, 3, 4\}$; 4
11. If the elements of the given set are placed in a one-to-one correspondence with the elements of an initial subset of the natural numbers, the last-named element in the correspondence names the cardinal number of the given set. This process is called counting the elements of the given set.
12. (a) Yes; their elements are in a one-to-one correspondence.   (c) Yes; since their elements may be placed in a one-to-one correspondence, they have the same cardinal number.

EXERCISE 2.8

1. Very likely. It is an extremely simple system and it is easy to invent and use.
2. (a) ∩∩∩∩||| (c) 𝓚 𝓚 ∞ ∞ ∞   𝑒𝑒𝑒 ∩∩∩
                                                  𝑒𝑒 ∩∩∩ ||
3. The two numbers are equal.
4. (a) 20,063   (c) 32,347
5.   ∩∩∩|||||
      𝓚 ∞ ∞ ſſ𝑒 ∩∩ ||||
6. (a) ▼▼▼▼   (c) ❮   ❮▼▼
7. The place value property.
8. (a) 24   (c) 782
9. (a) MCCXV   (c) MLXVI

10. DCXLIX, DCL, DCLI
11. CCXLVIII

EXERCISE 2.10

1. (a) Each of the ten basic symbols in our decimal system of notation is a digit.
   (b) The digits are 0, 1, 2, 3, 4, 5, 6, 7, 8, and 9.
2. A *numeral* is a name or symbol for a number.
3. (a) Because *decem* is the Latin word for ten and the value of each digit of a numeral is multiplied by some power of ten according to its position in the numeral. (b) The *value* of each digit is determined by its *place* (position) in the numeral. (c) The numeral for zero is used to properly position other digits with reference to the decimal point. Examples are 90.7, 0.0032, and 400.
4. The value of each digit in a numeral is multiplied by some power of ten according to its position in relation to the decimal point.
5. (a) 4376 (c) 0.0070809 (e) 806.0302
6. (a) $9 \cdot 10^2 + 4 \cdot 10^1 + 3 \cdot 10^0$ (c) $1 \cdot 10^2 + 0 \cdot 10^1 + 5 \cdot 10^0$ (e) $7 \cdot 10^{-3}$

EXERCISE 3.3

1. (a) 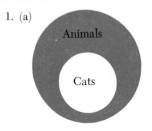 (c)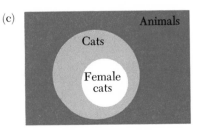

2. (a) The set of all male zebras. (c) The empty set.
3. (a) The set of all fish that are not trout. (c) The empty set.
4. (a) $\{d, e\}$ (c) $\{b, c, d\}$ (e) $\{a, b, c, d, e\}$
5.

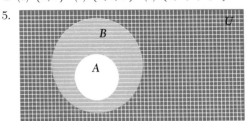

6. (a) C (c) D
7. (a) $\varnothing$

EXERCISE 3.5

1. (a) See Definition 3.4.1. (c) See Definition 3.5.2.
2. We mean the operation is performed on exactly two sets (which may be equal) and, in so doing, a third set is uniquely determined.

3. (a)

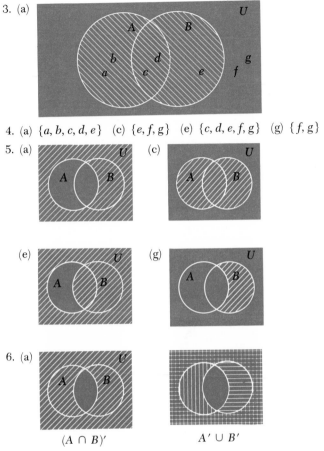

4. (a) $\{a, b, c, d, e\}$ (c) $\{e, f, g\}$ (e) $\{c, d, e, f, g\}$ (g) $\{f, g\}$

5. (a)   (c)

   (e)   (g)

6. (a)

$(A \cap B)'$ $A' \cup B'$

Since the shaded region of the left diagram corresponds to the union of the shaded regions of the right diagram, $(A \cap B)' = A' \cup B'$.

7. (a) The set of all girls with blond hair or blue eyes or both. (c) The set of all girls not having both blond hair and blue eyes. (e) The set of all girls having neither blond hair nor blue eyes.

8. (a) The set of all people who are redheaded or hot-tempered or both. (c). The set of all people who are neither redheaded nor hot-tempered. (e) The set of all people who are not redheaded or who are hot-tempered or who are not redheaded but are hot-tempered.

EXERCISE 3.6

1. (a) $A \cup B = \{a, b, c\} \cup \{c, d, e, f\} = \{a, b, c, d, e, f\}$
   $B \cup A = \{c, d, e, f\} \cup \{a, b, c\} = \{a, b, c, d, e, f\}$
   $A \cup B = B \cup A$ is verified.
   (c) $A \cap (B \cup C) = \{a, b, c\} \cap (\{c, d, e, f\} \cup \{a, e, f, g\})$
   $\qquad\qquad\qquad\quad = \{a, b, c\} \cap \{a, c, d, e, f, g\}$
   $\qquad\qquad\qquad\quad = \{a, c\}$

$$(A \cap B) \cup (A \cap C) = (\{a, b, c\} \cap \{c, d, e, f\}) \cup (\{a, b, c\} \cap \{a, e, f, g\})$$
$$= \{c\} \cup \{a\}$$
$$= \{a, c\}$$
$$A \cap (B \cup C) = (A \cap B) \cup (A \cap C) \text{ is verified.}$$

2. (a) A (c) Ø (e) A (g) Ø (i) B (k) Ø

3.

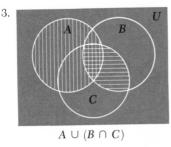

$A \cup (B \cap C)$

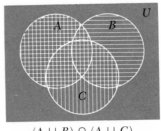

$(A \cup B) \cap (A \cup C)$

Since all shaded regions of the left diagram correspond to the crosshatched region of the right diagram, $A \cup (B \cap C) = (A \cup B) \cap (A \cup C)$ is verified.

4.

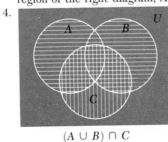

$(A \cup B) \cap C$

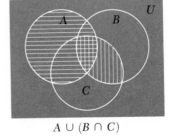

$A \cup (B \cap C)$

Since the crosshatched region of the left diagram does not correspond to the union of all shaded regions of the right diagram, $(A \cup B) \cap C \neq A \cup (B \cap C)$.

5.

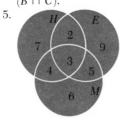

(a) 36 (b) 5 (c) 9 (d) 2 (e) 7

EXERCISE 3.7

1. (a) Yes. (c) Yes.
2. See Definition 3.7.1.
3. (a) $A \times B = \{(2, 7), (2, a), (3, 7), (3, a), (4, 7), (4, a)\}$ (c) 6 (e) No.
4. (a) $\{(4, 5), (4, 6), (4, 7), (5, 5), (5, 6), (5, 7), (6, 5), (6, 6), (6, 7)\}$ (c) 9 (e) 4
5. (a) 3 (c) Ø (e) 0
6. $A \times B = \emptyset$

EXERCISE 3.8

1.

C = Set of all clever people.
M = Set of all people who like mathematics.
I = Set of all interesting people.

(a) Valid.   (c) Valid.   (e) Valid.

2.

M = Set of all mathematics majors.
L = Set of all persons who like logic.
G = Set of all girls.

(a) Invalid.   (c) Valid.   (e) Invalid.

3.

B = Set of all beautiful people.
G = Set of all girls.
E = Set of all girls who like English.
H = Set of all girls who like history.

(a) Valid.   (c) Valid.   (e) Invalid.

EXERCISE 4.2

1. See Definition 4.2.1.
2. Because $A \cap B = \{b, c\}$ and $\{b, c\} \neq \varnothing$, so our definition for the sum of two whole numbers was violated.
3. (a) $n(A \cup B) = n(A) + n(B) - n(A \cap B)$ (b) Yes.
4. (a) For any two elements a and b in the whole numbers, there exists a unique whole number $a + b$ that we call the sum.   (b) There is a whole number that is their sum, and there is exactly one such number.
5. (a) Yes.   (b) Yes.   (c) No.   (d) Yes.
6. By counting we see that $n(A) = 2$ and $n(B) = 4$. Since $A \cap B = \varnothing$, by the definition of addition $2 + 4 = n(A \cup B) = 6$, and $4 + 2 = n(B \cup A) = 6$. Hence, $2 + 4 = 4 + 2$.
7. (a) For any elements a, b, and c in W, $(a + b) + c = a + (b + c)$.
8. As a counterexample: $8 \div 2 = 4$, but $2 \div 8 \neq 4$.
9. There exists a unique number 0 such that for any a in W, $a + 0 = 0 + a = a$.
10. (a) Commutative.   (c) Commutative.   (e) Associative and commutative.
    (g) Additive identity.

EXERCISE 4.4

1. (a) See Definition 4.3.1.   (b) 8
2. By Definition 4.3.1, $3 \times 0 = n(A \times \varnothing) = n(\varnothing) = 0$. Therefore, $3 \times 0 = 0$.
3. (a) For any two elements a and b in W, there exists a unique whole number $a \cdot b$ that we call the product of a and b.   (b) There is a whole number that is their product, and there is exactly one such number.
4. (a) Yes.   (b) Yes.   (c) Yes.   (d) Yes.   (e) Yes.   (f) No.
5. $a = 0$, or $b = 0$, or $a = 0$ and $b = 0$.

6. (b) No.
7. (a) For any elements a, b, and c in W, $(a \cdot b) \cdot c = a \cdot (b \cdot c)$.
8. (a) Commutative property of addition. (c) Commutative property of multiplication. (e) Commutative property of multiplication. (g) Commutative property of addition. (i) Associative property of addition.
9. There exists a unique number 1 in W such that for any a in W, $a \cdot 1 = 1 \cdot a = a$.
10. (a) $6(8 + 3) = 6(11) = 66$, and $6 \cdot 8 + 6 \cdot 3 = 48 + 18 = 66$. Hence, $6(8 + 3) = 6 \cdot 8 + 6 \cdot 3$.
11. [*Hint*: $34 = 30 + 4$.]
12. (a) $7a + 7n$ (c) $a(x + y)$ (e) $k^2(x + 1)$ (g) $a(x + y) + az = a[(x + y) + z] = a(x + y + z)$ (i) $a(x + 1)$
13. (a) $7x + 3x = (7 + 3)x = 10x$

EXERCISE 4.5

1. (a) See Definition 4.5.1. (c) $9 - 6 = 3$
2. Include new numbers in the mathematical system.
3. $8 - 3 = 5$, since $8 = 3 + 5$; but $3 - 8 \neq 5$, since $3 \neq 8 + 5$. Hence, $8 - 3 \neq 3 - 8$.
4. $(8 - 5) - 2 \neq 8 - (5 - 2)$, since $(8 - 5) - 2 = 3 - 2 = 1$, while $8 - (5 - 2) = 8 - 3 = 5$.
5. Example of left distributive property: $3(8 - 2) = 3 \cdot 8 - 3 \cdot 2$, since $3(8 - 2) = 3 \cdot 6 = 18$ and $3 \cdot 8 - 3 \cdot 2 = 24 - 6 = 18$. Using the right distributive property: $(8 - 2)3 = 8 \cdot 3 - 2 \cdot 3$, since $(8 - 2)3 = 6 \cdot 3 = 18$ and $8 \cdot 3 - 2 \cdot 3 = 24 - 6 = 18$.
6. (a) See Definition 4.5.2. (c) If $0 \div 0 = q$, it is necessary by the definition of division that $0 = 0 \cdot q$. But all values of q satisfy this equation and q is not unique. Hence, we bar division by zero.
7. None of the given properties hold.
8. (a) Yes. (b) No.
9. $(12 + 16) \div 4 = (12 \div 4) + (16 \div 4)$. There are many other examples.
10. $(16 - 8) \div 2 = (16 \div 2) - (8 \div 2)$. There are many other examples.

EXERCISE 4.6

	Set S	$n(S)$
1.	$\{[a \quad b \quad c \quad d \quad e \quad f \quad g \quad h \quad i \quad j] \, [k \quad l \quad m \quad n \quad o]\}$	15_{ten}
	$\{[a \quad b \quad c \quad d \quad e \quad f \quad g] \, [h \quad i \quad j \quad k \quad l \quad m \quad n]o\}$	21_{seven}

	Set S	$n(S)$
2.	$\{\, [a \quad b \quad c \quad d] \, [e \quad f \quad g \quad h] \quad [i \quad j \quad k \quad l] \quad m \quad n\}$	32_{four}
	$\{[[a \quad b \quad c] \, [d \quad e \quad f] \, [g \quad h \quad i]] \, [j \quad k \quad l] \quad m \quad n\}$	112_{three}

3. (a) Seven. (c) k
4. (a) Four sevens and three. (c) Three seven-sevens, zero sevens, and four.
5. 1, 2, 3, 4, 5, 6, 10, 11, 12, 13, 14, 15, 16, 20, 21
6. In base six we would not have the digit 7.
7. $3_{\text{four}} = 3_{\text{five}}$, since the digits have the same place value; but $12_{\text{four}} \neq 12_{\text{five}}$, since the digits of 1 have different place values.
8. (a) 2 (c) 29
9. (a) 11_{two} (c) 1100011_{two}
10. (a) 15_{twelve} (c) TE_{twelve}

11. (a) 19   (c) 99
12. (a) 14_{seven} (c) 131_{seven}

EXERCISE 4.7

1. (a) 11   (c) 68
2.

+	0	1	2	3	4	5	6
0	0	1	2	3	4	5	6
1	1	2	3	4	5	6	10
2	2	3	4	5	6	10	11
3	3	4	5	6	10	11	12
4	4	5	6	10	11	12	13
5	5	6	10	11	12	13	14
6	6	10	11	12	13	14	15

3. (a) 10   (c) 2001
4. (a) 2   (c) 125
5. (a) Three.   (c) Fourteen.   (e) Two.
6. (a) Four.   (c) Nine.
7. (a) 11   (c) 1011   (e) 110100
8. (a) 9T   (c) 740   (e) 1819

EXERCISE 4.8

1.

·	0	1
0	0	0
1	0	1

2.

·	0	1	2	3	4
0	0	0	0	0	0
1	0	1	2	3	4
2	0	2	4	11	13
3	0	3	11	14	22
4	0	4	13	22	31

3. (a) No.   (c) Although the multiplier 23_{four} is equal to 11_{ten}, the partial products are different because $3_{\text{four}} \neq 1_{\text{ten}}$ and $20_{\text{four}} \neq 10_{\text{ten}}$.
4. (a) 4   (c) 234   (e) 4141
5. (a) Five.   (c) Three.   (e) Eight.

6. (a) 12 (c) 10 (e) 6 (g) 6
7. (a) 11. *Check:* $1 \times 3 = 3$. (c) 10110. *Check:* $2 \times 11 = 22$.
8. (a) The new numeral represents the original number multiplied by ten.
   (c) The new numeral represents the original multiplied by seven.
   (e) $23_{\text{four}} = 11$; $230_{\text{four}} = 44$; and $4 \times 11 = 44$.
9. (a) $q = 23$. *Check:* $39 \div 3 = 13$. (c) $q = 24$. *Check:* $98 \div 7 = 14$.
10. (a) Checking base four: $30 \div 6 = 5$. Checking base five: $42 \div 7 = 6$.
    (c) Because 3 is used as a digit, and the only digits in base three are 0, 1,
    and 2.
11. Base six.

EXERCISE 5.3

1. A number system consists of a set of elements called numbers and of operations
   defined on these numbers.
2. (a) Addition and multiplication.
3. See Definition 5.2.1.
4. (a) -5
5. $J = \{ \ldots, -3, -2, -1, 0, 1, 2, 3, \ldots \}$
6. (a) When the sum of two integers, $a + b$, is equal to zero, b is called the
   additive inverse of a. (c) 7
7. (a) -3 (c) -7
8. (a) No. Zero is neither a positive number nor a negative number. (c) -0
   and 0 are numerals for the same number and are therefore equal; $-0 = 0$.

EXERCISE 5.5

1. Each element in J has an additive inverse.
2. (a) -3 (c) 13 (e) -1
3. (a) $-6 + 14 = -6 + (6 + 8) = (-6 + 6) + 8 = 0 + 8 = 8$.   Therefore,
   $-6 + 14 = 8$.
4. (a) -5 (c) -15 (e) -89
5. 14,777 feet
6. (a) -2 (c) 8 (e) -23 (g) 7
7. $6 - (-4) = (6 + 4) = 10$, while $-4 - 6 = -10$. Therefore, $6 - (-4) \neq$
   $-4 - 6$.
8. $(-8 - 4) - 6 = -12 - 6 = -18$, while $-8 - (4 - 6) = -8 - (-2) =$
   $-8 + 2 = -6$. Hence, $(-8 - 4) - 6 \neq -8 - (4 - 6)$.
9. (a) -6 (c) -15 (e) -12 (g) 0
10. (a) -8 (c) 12
11. See the solution to a similar problem in Section 5.4.
12. (a) -66
13. (a) -4 (c) 0 (e) 8 (g) 0 (i) 1
14. $2 \div 6$ is not an integer since there is no integer q such that $2 = 6q$.
15. Either $a = b$ or $a = -b$, $a \neq 0$ and $b \neq 0$.

EXERCISE 5.6

1. See Definition 5.6.1.
2. See Definition 5.6.2.
3. Examples are $\frac{10}{6}$, $\frac{-15}{-9}$, and $\frac{20}{12}$
4. Four.

5.

6. $\frac{11}{4}$, since $2\frac{3}{4} = 2 + \frac{3}{4} = \frac{8}{4} + \frac{3}{4} = \frac{11}{4}$.

7. See Definition 5.6.3.

8. (a) $\frac{1}{2}$ (c) $\frac{-2}{3}$ (e) $\frac{-2}{3}$ (g) $\frac{3}{1}$ (i) $\frac{0}{1}$

9. $\frac{0}{2}, \frac{0}{3}, \frac{0}{4}, \frac{0}{5}$, and $\frac{0}{6}$

10. No. The denominator may not be 0.

11. No. $\frac{2}{4}$ is not in simplest form; hence elements such as $\frac{1}{2}, \frac{3}{6}$, and $\frac{5}{10}$ will not be included.

EXERCISE 5.8

1. (a) $\frac{17}{12}$ (c) $\frac{25}{9}$ (e) $\frac{-25}{24}$

2. $\frac{2}{7} + \frac{4}{5} = \frac{10+28}{35} = \frac{38}{35}$, and $\frac{4}{5} + \frac{2}{7} = \frac{28+10}{35} = \frac{38}{35}$. Hence, $\frac{2}{7} + \frac{4}{5} = \frac{4}{5} + \frac{2}{7}$.

3. (a) 0 (c) $\frac{2}{3}$

4. (a) Because the greatest common integral divisor of 0 and 1 is 1, and the denominator 1 is positive.

5. (a) $\frac{-2}{5}$ (c) $\frac{-7}{2}$ (e) $\frac{2}{7}$

6. (a) $\frac{1}{12}$ (c) $\frac{-43}{8}$ (e) $\frac{65}{24}$

7. (a) $\frac{3}{4} - \frac{1}{2} = \frac{1}{4}$, while $\frac{1}{2} - \frac{3}{4} = \frac{-1}{4}$. Since $\frac{1}{4} \neq \frac{-1}{4}, \frac{3}{4} - \frac{1}{2} \neq \frac{1}{2} - \frac{3}{4}$.

8. (a) $\frac{3}{7} \cdot \frac{4}{5} = \frac{12}{35}$ and $\frac{4}{5} \cdot \frac{3}{7} = \frac{12}{35}$. Hence, $\frac{3}{7} \cdot \frac{4}{5} = \frac{4}{5} \cdot \frac{3}{7}$.

9. (a) $\frac{4}{3}$ (c) $\frac{3}{2}$ (e) $\frac{-5}{6}$

10. (a) $\frac{3}{10}$ (c) $\frac{1}{1}$ (e) $\frac{28}{3}$

11. $(\frac{3}{5} \div \frac{2}{3}) \div \frac{7}{2} = \frac{9}{35}$, while $\frac{3}{5} \div (\frac{2}{3} \div \frac{7}{2}) = \frac{63}{20}$. Hence, $(\frac{3}{5} \div \frac{2}{3}) \div \frac{7}{2} \neq \frac{3}{5} \div (\frac{2}{3} \div \frac{7}{2})$.

12. $\frac{3}{8}$

13. $\frac{a}{c} + \frac{b}{c} = \frac{ac+cb}{c^2} = \frac{c(a+b)}{c \cdot c} = \frac{c}{c} \cdot \frac{a+b}{c} = \frac{a+b}{c}$.

14. (a) c may be any positive or negative integer.

EXERCISE 5.9

1. (a) $7 < 10$ since $7 + 3 = 10$ and 3 is positive. (c) $-14 < 6$ since $-14 + 20 = 6$ and 20 is positive.

2. (a) True. (c) False. (e) True. (g) True.

3. (a) $x = 9$ (c) $\{x \mid x \in J \text{ and } x > 9\}$

4. (a) $\frac{3}{4} < \frac{4}{5}$ (c) $\frac{-3}{4} < \frac{2}{3}$ (e) $\frac{16}{28} < \frac{11}{18}$

5. Between any two distinct rational numbers we can always find another rational number and hence infinitely many rational numbers.

6. (a) $\frac{19}{36}$ (c) $\frac{19}{5}$

7. (a) Five: $-3, -2, -1, 0$, and 1. (c) 0

8. (a) Infinitely many. $\{x \mid x \in R \text{ and } \frac{3}{1} < x < \frac{4}{1}\}$. Two examples are $\frac{7}{2}$ and $\frac{11}{3}$.
   (c) Infinitely many. $\{x \mid x \in R \text{ and } \frac{3}{200} < x < \frac{4}{200}\}$. Two examples are $\frac{1}{60}$ and $\frac{11}{600}$.

EXERCISE 5.11

1. Proof is given in the text.

2. $c^2 = 4(\frac{1}{2}ab) + x^2$
   $c^2 = 2ab + (b - a)^2$
   $c^2 = 2ab + b^2 - 2ab + a^2$
   $c^2 = b^2 + a^2$
   $a^2 + b^2 = c^2$

3. (a) 12 feet.
4. $\sqrt{50}$ units or $5\sqrt{2}$ units.
5. Proof is in the text.
6. Assume $\sqrt{3}$ is a rational number indicated by $\frac{a}{b}$ in simplest form. Then $3 = \frac{a^2}{b^2}$ and $3b^2 = a^2$. Since $3b^2$ has a factor of 3, a^2 must have a factor of 3. Hence, a must have a factor of 3 since 3 is prime, and we may let $a = 3k$. Then $3b^2 = (3k)^2$, or $3b^2 = 9k^2$. Thus, $b^2 = 3k^2$. By again reasoning as above, b^2 must have a factor of 3 and hence b has a factor of 3. But if both a and b have a factor of 3, our assumption that $\frac{a}{b}$ is in simplest form is contradicted. Since any rational number can be represented by a fraction in simplest form, $\sqrt{3}$ must be irrational.
7. (a) Assume $7 + \sqrt{2} = r_1$, a rational number. Then $\sqrt{2} = r_1 - 7$. But the rationals have closure for subtraction; hence, we conclude that $r_1 - 7$ is a rational number r_2. Then $\sqrt{2} = r_2$, a rational number. However, since we know $\sqrt{2}$ is irrational, our assumption that $7 + \sqrt{2}$ is rational must be false.
8. (a) 16. No.   (b) $-2\sqrt{3}$. Yes.   (c) 61. No.

EXERCISE 5.14

1. (a) $0.\overline{21}$ (c) 0.125 (e) $0.\overline{30}$ (g) $0.\overline{714285}$ (i) $0.2\overline{52}$
2. (a) $\frac{4}{9}$ (c) $\frac{54}{99} = \frac{6}{11}$ (e) $\frac{393}{999} = \frac{131}{333}$ (g) $\frac{889}{9}$ (i) $\frac{101}{999}$
3. (a) Twelve: $1, 2, \ldots, 12$.   (c) $0.\overline{384615}$
4. The simplest form of $\frac{20}{52}$ is $\frac{5}{13}$. Hence, the maximum possible number of digits in the repeating portion of the quotient is $13 - 1 = 12$.

5. (a)   Let $x = 0.44\overline{9}$ (c)   Let $x = 0.\overline{9}$
         Then $10x = 4.49\overline{9}$ Then $10x = 9.\overline{9}$
         Subtract $x = 0.44\overline{9}$ Subtract $x = 0.\overline{9}$
         ───────────   ───────────
              $9x = 4.05$ $9x = 9$
              $x = 0.45$ $x = 1$

6. $\frac{abcd}{9999}$
7. (a) 2.6

EXERCISE 5.15

1. (a) -1 (b) $-\sqrt{-1}$ (c) 1 (d) $\sqrt{-1}$ (e) -1 (f) Yes.
2. (a) $x = \pm 7$ (c) $x = \pm 7i$ (e) $x = \pm 12i$
3. (a) Newton used the works of others as the foundation for his work.
4. (a) $9 + 7i$ (c) $2 - i$ (e) $18 - i$
5. The given set has the same cardinal number as N since their elements can be placed in a 1–1 correspondence:

$$\{1, 2, 3, 4, \ldots, n, \ldots\}$$
$$\updownarrow \updownarrow \updownarrow \updownarrow \qquad \updownarrow$$
$$\{1, 4, 9, 16, \ldots, n^2, \ldots\}$$

6. Since the odd numbers may be placed in a 1–1 correspondence with the natural numbers, they have the same cardinal number $\aleph_0$:

$$\{1, 2, 3, 4, \ldots, \quad n, \quad \ldots\}$$
$$\updownarrow \updownarrow \updownarrow \updownarrow \qquad \updownarrow$$
$$\{1, 3, 5, 7, \ldots, 2n-1, \ldots\}$$

7. This problem is analogous to illustrative Example 1 of Section 5.15.

8 This problem is similar to illustrative Example 1 of Section 5.15.

9. Show that $A = \{1, 3, 5, \ldots, 2n - 1, \ldots\}$ and $B = \{2, 4, 6, \ldots, 2n, \ldots\}$ each have cardinal number $\aleph_0$. Since $A \cap B = \varnothing$, $n(A \cup B) = \aleph_0 + \aleph_0$. But $A \cup B = N$, which has cardinal number $\aleph_0$; hence $\aleph_0 + \aleph_0 = \aleph_0$.

EXERCISE 6.2

1. See Definition 6.2.1.
2. See Definition 6.2.2.
3. (a) $\{2, 3, 5, 7, 11, 13, 17, 19, 23, 29, 31, 37, 41, 43, 47, 53, 59, 61, 67, 71, 73, 79,$ $83, 89, 97\}$
4. $6 = 3 + 3$; $8 = 3 + 5$; $10 = 3 + 7 = 5 + 5$; $12 = 5 + 7$; $72 = 5 + 67$ or $11 + 61$ or $13 + 59$ or $19 + 53$ or $29 + 43$ or $31 + 41$; $84 = 5 + 79$ or $11 + 73$ or $13 + 71$ or $17 + 67$ or $23 + 61$ or $31 + 53$ or $37 + 47$ or $41 + 43$.
5. $22 = 3 + 19 = 5 + 17 = 11 + 11$.
6. $9 = 3 + 3 + 3$; $11 = 3 + 3 + 5$; $13 = 3 + 3 + 7 = 3 + 5 + 5$; $15 = 3 + 5 + 7 = 5 + 5 + 5$; $35 = 3 + 3 + 29 = 3 + 13 + 19 = 5 + 7 + 23 = 5 + 11 + 19 = 5 + 13 + 17 = 7 + 11 + 17 = 11 + 11 + 13$.
7. 3, 5, 17, and 257.
8. (a) 40322, 40323, 40324, 40325, 40326, 40327, and 40328.

EXERCISE 6.3

1. (a) $60 = 2^2 \cdot 3 \cdot 5$ (c) $51 = 3 \cdot 17$ (e) $510 = 2 \cdot 3 \cdot 5 \cdot 17$
2. Only one prime factorization is possible: $2^3 \cdot 3$.
3. $-18 = -1 \cdot 2 \cdot 3^2$
4. $28 = 1 + 2 + 4 + 7 + 14$.
5. $9^3 + 10^3 = 1729$.

EXERCISE 6.4

1. Let a and b be members of W, then $(2a + 1)$ and $(2b + 1)$ represent any two odd numbers. The sum $(2a + 1) + (2b + 1) = 2a + 2b + 1 + 1 = 2a + 2b + 2 = 2(a + b + 1) = 2k$, where $k \in W$. But $2k$ is an even number. Hence the sum of any two odd numbers is an even number.
2. If a and b are in W, then $2a$ represents any even number and $(2b + 1)$ any odd number. The sum $2a + (2b + 1) = (2a + 2b) + 1 = 2(a + b) + 1 = (2k + 1)$, where $k \in W$. But $(2k + 1)$ is an odd number. Hence the sum of any even number and any odd number is an odd number.
3. If a and b are in W, then $(2a + 1)$ and $(2b + 1)$ represent any two odd numbers. If the product $(2a + 1)(2b + 1)$ exists and is unique, the set of odd numbers is closed under multiplication. But $(2a + 1)(2b + 1) = (2a + 1) \times (2b) + (2a + 1)(1) = 4ab + 2b + 2a + 1 = 2(2ab + b + a) + 1 = (2k + 1)$, where $k \in W$. Since $(2k + 1)$ exists and is a unique odd number, the set of odd numbers is closed under multiplication.
4. If a and b are in W, then $2a$ represents any even number and $(2b + 1)$ any odd number. The product $(2a)(2b + 1) = 4ab + 2a = 2(2ab + a) = 2k$, where $k \in W$. But $2k$ is an even number. Hence the product of an even number and an odd number is an even number.
5. (a) $3 \cdot 5 + 1 = 16 = 4^2$; $5 \cdot 7 + 1 = 36 = 6^2$; $11 \cdot 13 + 1 = 144 = 12^2$.
6. (a) $5^2 = 25$ (b) $1 + 3 + 5 + \cdots + (2n - 1) = n^2$ where n is any natural number. For example: $1 + 3 + 5 + 7 + 9 + 11 + 13 = 7^2 = 49$.
7. If a is a factor of some number n, then there must exist a number b such that $a \cdot b = n$. Hence factors occur in pairs, and there will be an even number

of distinct factors of any given number n unless there is some factor c such that $c \cdot c = n$. In this case, there will be $(2k + 1)$ or an odd number of distinct factors of n. But if $c \cdot c = n$, n is a perfect square of c, and we see that a number with an odd number of distinct factors is a perfect square.

EXERCISE 6.6

1. A number is *perfect* if it is equal to the sum of its proper divisors.
2. 6, 28, 496, and 8128.
3. (a) No one knows (probably infinitely many). (b) No one knows (probably not).
4. (a) (This is a matter of opinion.) (b) (Who knows!) (c) (The author would not since some people believe 13 to be an unlucky number.) (d) (The author refuses to answer this question!)
5. The sum of the proper divisors of a is equal to b, and the sum of the proper divisors of b is equal to a.
6. (a) 220 and 284; next in order are 1184 and 1210.
7. (a) $\{3, 4, 5\}$, $\{5, 12, 13\}$, $\{7, 24, 25\}$, and $\{9, 40, 41\}$.
8. 25 coconuts were collected.
9. There is more than one possible solution. A six-step solution showing the number of gallons in each jug after each step follows:

8-gal. jug	5-gal. jug	3-gal. jug
8	0	0
3	5	0
3	2	3
6	2	0
6	0	2
1	5	2
1	(4)	3

EXERCISE 7.2

1. (a)

+	0	1	2
0	0	1	2
1	1	2	0
2	2	0	1

(c) Addition is associative.

2. (a)

+	0	1	2	3	4
0	0	1	2	3	4
1	1	2	3	4	0
2	2	3	4	0	1
3	3	4	0	1	2
4	4	0	1	2	3

(c) 0   (e) 12

3. In effect, a 24-hour clock is being used instead of a 12-hour clock. This avoids possible errors inherent in using the a.m. and p.m. symbols. The expression 0300 hours would mean 3 a.m. The expression 1620 hours means 4:20 p.m.
4. 9 a.m.
5. 1 o'clock.

EXERCISE 7.3

1. See Definition 7.3.1.
2. The commutative property need not necessarily hold for the binary operation of a given group. If it does, however, the system is a commutative group.
3. See Definition 7.3.2.
4. See Definition 7.3.3.
5. (a) The four-hour clock system is a group under addition but not under multiplication. (c) The four-hour clock system is a ring under the operations of addition and multiplication.
6. (a) The five-hour clock system is a group under addition but not under multiplication. (c) The five-hour clock system is a ring under the operations of addition and multiplication.
7. (a) The element 0 must be removed from the set since it has no multiplicative inverse.
8. (a) $a - b = k$ if and only if $a = b + k$. (c) 4
9. The system of integers is a ring but not a field since the property concerning multiplicative inverses is not satisfied.
10. The system of rational numbers is both a ring and a field.

EXERCISE 7.4

1. See Definition 7.4.1.
2. $46 \equiv 22 (\text{mod } 6)$, since $46 - 22 = 24 = 4(6)$.
3. (a) 1 (c) 2 (e) 2 (g) 2 (i) 1
4. (a) 2 (b) $20 \equiv 2 (\text{mod } 3)$, since $20 - 2 = 18 = 6(3)$.
5. (a) $3 (\text{mod } 5)$ (c) $1 (\text{mod } 6)$ (e) $8 (\text{mod } 9)$ (g) $6 (\text{mod } 7)$ (i) $1 (\text{mod } 9)$
6. (a)

+	0	1	2	3	4	5	6
0	0	1	2	3	4	5	6
1	1	2	3	4	5	6	0
2	2	3	4	5	6	0	1
3	3	4	5	6	0	1	2
4	4	5	6	0	1	2	3
5	5	6	0	1	2	3	4
6	6	0	1	2	3	4	5

(c) Yes. Every property necessary to qualify as a commutative group is satisfied. (e) Yes.

EXERCISE 7.5

1. (a) $23 \equiv 5 (\text{mod } 9)$ (c) $23,425 \equiv 7 (\text{mod } 9)$ (e) $823,429 \equiv 1 (\text{mod } 9)$
2. (a) 1462 (c) 144,692

3. (a) 17,782   (c) 876,042
4. (a) $q = 18$ and $r = 12$ (c) $q = 1857$ and $r = 2$
5. Treat the minuend as the sum of the subtrahend and the remainder. Then proceed as in checking addition by casting out nines.

EXERCISE 8.2

1. 15
2. (a) 1   (c) 720   (e) 32
3. (a) 12
4. 60
5. $2^8 = 256$
6. $2^{10} = 1024$
7. 303,600
8. (a) $5! = 120$
9. $2^6 - 1 = 63$
10. (a) $26 \cdot 25 \cdot 24 \cdot 23 = 358,800$ (c) $3 \cdot 26^3 = 52,728$
11. 400

EXERCISE 8.3

1. (a) 1   (c) 20   (e) 120
2. $6! = 720$
3. (a) $3! = 6$
4. (a) $6! = 720$
5. $18 \cdot 10^8 = 1,800,000,000$
6. $(6!)(4!) = 17,280$
7. $\frac{7!}{3!\,3!} = 140$
8. $\frac{9!}{2!\,3!\,4!} = 1260$
9. 720
10. (a) $7! = 5040$
11. (a) $2(4!)(3!) = 288$ (b) $4!4! = 576$

EXERCISE 8.4

1. $C(n, 0) = \dfrac{P(n, 0)}{0!} = \dfrac{1}{1} = 1$
2. (a) 1   (c) 10   (e) 5
3. $\frac{10!}{5!\,5!} = 252$
4. $\frac{10!}{7!\,3!} = 120$
5. (a) $\frac{5!}{2!\,3!} = 10$ (c) 1
6. $\frac{10!}{7!\,3!} = 120$
7. $\frac{2(98!)}{94!\,4!} = 7,224,560$
8. (a) $[C(13, 3)] \cdot [C(39, 2)] = \frac{13!}{10!\,3!} \cdot \frac{39!}{37!\,2!} = 211,926$
   (b) $[C(13, 3)] \cdot [C(39, 2)] + [C(13, 4)] \cdot [C(39, 1)] + C(13, 5)$
   $= \frac{13!}{10!\,3!} \cdot \frac{39!}{37!\,2!} + \frac{13!}{9!\,4!} \cdot \frac{39!}{38!\,1!} + \frac{13!}{8!\,5!} = 241,098$
   (c) $C(39, 5) + [C(13, 1)] \cdot [C(39, 4)]$
   $= \frac{39!}{34!\,5!} + \frac{13!}{12!\,1!} \cdot \frac{39!}{35!\,4!} = 1,645,020$

EXERCISE 8.6

1. (a) $\frac{1}{4}$ (c) $\frac{3}{4}$ (e) 1
2. (a) 5

3. (a) $\frac{1}{18}$

4. (a) $\dfrac{C(6,5)}{C(10,5)} = \dfrac{1}{42}$ (c) $\dfrac{C(6,3)C(4,2)}{C(10,5)} = \dfrac{10}{21}$

5. (a) $\frac{1}{6}$ (c) 1 to 5

6. (a) $\frac{9}{19}$ (c) 9 to 10

7. (a) $\dfrac{C(26,2)}{C(52,2)} = \dfrac{25}{102}$ (c) $\dfrac{C(4,2)}{C(52,2)} = \dfrac{1}{221}$

8. (a) $\dfrac{C(5,2)C(3,2)}{C(8,4)} = \dfrac{3}{7}$

EXERCISE 8.7

1. (a) $\frac{1}{6}$ (c) $\$.16\overline{6}$ or $16\frac{2}{3}$¢
2. (a) \$2
3. \$9
4. (a) $3\frac{1}{2}$ (c) (It will likely be close to 70.)
5. (a) \$1   (c) \$100   (e) \$1000

EXERCISE 8.8

1. (a) $C(7,5) = 21$
2. $C(7,3) = \frac{7 \cdot 6 \cdot 5}{3!} = 35$, and $C(7,4) = \frac{7 \cdot 6 \cdot 5 \cdot 4}{4!} = 35$.
3. $\{1, 5, 10, 10, 5, 1\}$
4. (a) $2^8 = 256$
5. (a) $2^6 = 64$ (c) $\dfrac{C(6,3)}{2^6} = \dfrac{5}{16}$
6. (a) $2^7 = 128$ (c) $\dfrac{C(7,5)}{2^7} = \dfrac{21}{128}$
7. $\{1, 7, 21, 35, 35, 21, 7, 1\}$, and $\{1, 8, 28, 56, 70, 56, 28, 8, 1\}$.
8. $\{C(5, 0),\ C(5, 1),\ C(5, 2),\ C(5, 3),\ C(5, 4),\ C(5, 5)\}$; $\{C(6, 0),\ C(6, 1),\ C(6, 2),$ $C(6, 3),\ C(6, 4),\ C(6, 5),\ C(6, 6)\}$; $\{C(7, 0),\ C(7, 1),\ C(7, 2),\ C(7, 3),\ C(7, 4),$ $C(7, 5),\ C(7, 6),\ C(7, 7)\}$

EXERCISE 8.9

1. (a) $a^4 + 4a^3b + 6a^2b^2 + 4ab^3 + b^4$
   (a) $125x^3 + 75x^2y^2 + 15xy^4 + y^6$
2. (a) 9   (c) $C(8, 3) = 56$
3. (a) $C(6, 2) = 15$
4. $2^4 = 16$,   and $(1 + 1)^4 = C(4, 0)(1)^4 + C(4, 1)(1^3)(1) + C(4, 2)(1^2)(1^2) +$ $C(4, 3)(1)(1^3) + C(4, 4)(1^4) = 1 + 4 + 6 + 4 + 1 = 16$.
5. (a) $1 + 0.2 + 0.018 + 0.00096 = 0.21896$
   (c) Use more terms of the expansion of $(1 + .02)^{10}$.

EXERCISE 9.1

1. (a) and (e); (b) and (d); (c) and (f).
2. (a) and (f); (b) and (d); (c) and (e).
3. (a) and (f); (b) and (d); (c) and (e).
4. (a) Starting at the top of the diagram, the line segments cross respectively: 0, 1, 2, 3, 4, and 5 times.   (b) Yes.   (c) In the first instance choose the word *even*, and in the second instance choose the word *odd*.
5. (a) 24 inches.   (c) 9; 12.

6. (a) 2 (b) 3 (c) 4; 5; $n + 2$.
7. (a) No. (b) Yes.
8. (a) Yes. (b) No. (c) No. (d) A doughnut and a coffee cup are topologically equivalent.
9. (a) Yes. (c) Yes. (e) No. (g) No.

EXERCISE 9.2

1. Those of (a) and (b).
2. (a) A, B, C, or D. (c) The network of (c) is not traversable.
3. LM, MQ, QP, or PL.
4. (a) Yes, since there is only one pair of odd vertices.
5. (a) 8 (c) 8
6. (a) Two, since there are two pairs of odd vertices. (b) Only one; an additional bridge (edge) between any two vertices in Figure 9.2.2 will result in only one pair of odd vertices. (c) Two; by making a new bridge (edge) between any two points in Figure 9.2.2, and then another bridge between the remaining two points, all vertices will be even.

7. (a) Yes;  (b) No.

EXERCISE 9.3

1. (a) 13 (c) 24
2. Those of (a), (c), and (d).
3. (a) 2 (c) Yes: $2 + 3 - 4 = 1$. (e) Yes, since edges of a network may, by definition, intersect only at their end points.
4. (a) 5 (c) No. We know $F + V - E = 1$; but if $F = V$, then $2F - E = 1$ or $2F = E + 1$. Since $2F$ is even, $E + 1$ must be even. But $E + 1$ is even only if E is odd.
5. (a) Yes. (c) n
6. (a) 8 (c) 12 (e) Yes, all vertices are even.
7. (a) Two, since there are 2 pairs of odd vertices. (c) $3 + 4 - 6 = 1$.

EXERCISE 9.4

1. (a) 1 (c) 2
2. Those of (a) and (d) are connected networks.
3. (a) 3 colors. (b) 3 colors. (c) 3 colors. (d) 3 colors.
4. (a) 4 colors. (b) 4 colors.
5. (a) Yes. (c) Any map with only even interior vertices can be colored with only 2 colors.
6. (a)  (c)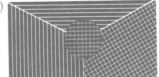

7. No.
8. (a) Yes. (c) Even.

EXERCISE 9.5

1. (a) Yes. (c) Yes. (e) One.
2. (a) Two. (c) Two. (e) Four. (g) Two.
3. (a) Two.
4. (a) Two.
5. (a) It has a "knot" in it.
6. (a) Two sides and three edges. (c) One side and one edge.

EXERCISE 10.2

1. By building tools, machines, and so on.
2. It can make computations rapidly and accurately, and print out the results.
3. Analog computers: thermometers, bathroom scales, watches. Digital computers: desk calculators, cash registers, abacuses.
4. (a) See Figure 10.2.1 for assistance. (c) An analog device, since the principle of measurement is used.

EXERCISE 10.3

1. (a) Digital.
2. (a) Analog. (c) Analog. (e) Analog. (g) Digital. (i) Digital.
3. See discussion of the text or other references.
4. (a) Technology had not advanced far enough to build the computer he designed.

EXERCISE 10.4

1. See discussion of the text or other references.
2. The two symbols of base two (0 and 1) can easily be represented by an open or closed circuit, the presence or absence of a hole in a card, magnetizing a core in either of two directions, and so on.
3. Much mechanical motion was eliminated from the computer by the use of vacuum tubes.
4. See discussion of the text or other references.
5. As compared with vacuum tubes, transistors are smaller, create less heat, are more economical to build and operate, and are highly reliable.
6. John von Neumann did much to develop the concept of a stored program.
7. The computer's "memory" refers to its ability to store operating instructions and other data.
8. 21.04941369 . . . (The time in computing this result will vary.)

EXERCISE 10.6

1–4. See discussion of the text or other references. (No unique answer is expected.)

INDEX

LIST OF SYMBOLS

Symbol	Meaning
$P \wedge Q$	P and Q
$P \vee Q$	P or Q
$\sim P$	Not P
$P \rightarrow Q$	If P, then Q (or P implies Q)
$P \leftrightarrow Q$	P if and only if Q (or P is equivalent to Q)
$\therefore$	Therefore
$\{a, b, c, \dots\}$	The set whose elements are a, b, c, and so on
$\{x \mid x \text{ has a certain property}\}$	The set of all x such that x has a certain property
$=$	Is equal to
$\neq$	Is not equal to
$\approx$	Is approximately equal to
$x \in A$	x is an element of set A
$x \notin A$	x is not an element of set A
$\emptyset$	The empty set
$A \subseteq B$	A is a subset of B
$A \subset B$	A is a proper subset of B
A'	The complement of set A
$n(A)$	The cardinal number of set A
$N = \{1, 2, 3, \dots\}$	The set of natural numbers
$W = \{0, 1, 2, 3, \dots\}$	The set of whole numbers
$J = \{\dots, -2, -1, 0, 1, 2, \dots\}$	The set of integers